Airline Maintenance Resource Management: Improving Communication

J.C. Taylor

and

T.D. Christensen

Society of Automotive Engineers, Inc.
Warrendale, Pa.

Library of Congress Cataloging-in-Publication Data

Taylor, James C. (James Chapman), 1937–
  Airline maintenance resource management : improving communication / J.C. Taylor and T.D. Christensen.
    p.   cm.
  "SAE order no. R–192"—T.p. verso.
  Includes bibliographical references and index.
  ISBN 0-7680-0231-1
  1. Transport planes—Maintenance and repair—Quality control.
  2. Aeronautics—Safety measures.   3. Airlines—Management.
  4. Business communication.   I. Christensen. T.D. (Tom D.)
  II. Title.
  TL671.9.T37   1998   98-36672
  387.7'3340423—dc21   CIP

Copyright © 1998 Society of Automotive Engineers, Inc.
        400 Commonwealth Drive
        Warrendale, PA 15096-0001 U.S.A.
        Phone: (724) 776-4841
        Fax: (724) 776-5760
        E-mail: publications@sae.org
        http://www.sae.org

ISBN 0-7680-0231-1

All rights reserved. Printed in the United States of America.

Permission to photocopy for internal or personal use, or the internal or personal use of specific clients, is granted by SAE for libraries and other users registered with the Copyright Clearance Center (CCC), provided that the base fee of $.50 per page is paid directly to CCC, 222 Rosewood Dr., Danvers, MA 01923. Special requests should be addressed to the SAE Publications Group. 0-7680-0231-1/98-$.50.

SAE Order No. R-192

*Dedicated to the airline mechanics,
engineers, and maintenance managers—
those proud professionals—
who keep the commercial fleet flying safely.*

Acknowledgments

We owe a great deal to several friends and benefactors, Bruce Aubin and John Goglia (NTSB), for the support and patience they offered while our ideas were turning into a readable work. Thanks go to all of them—without their guidance and foresight we would have had a very different product.

We wish to extend our appreciation to the many people who read various sections of the book during its creation. Among these are Jerry Allen, Phil Chartrand, Steve Erickson, Leslie Eveland, Wayne Gallimore, Jon Glynn, Jim Haarbauer, Jack Hessburg, Joe Kania, Jeff Leeds, Steve Miller, Manoj Patankar, Sam Sexhus, John Stelly Jr., and Larry Strouse. We also want to thank Gordon Dupont, Steve Predmore, Giselle Richardson, and William Whitley for generously allowing us to use their material, which we adapted and used in Chapters 2 and 11.

The book is based on research that began in 1990 and is ongoing. This work has been given financial and/or moral support by many people: Barbara Kanki (NASA Ames Research Center), Jean Watson (FAA Office of Aviation Medicine), Fred Leonelli (FAA Flights Standards), Chris Seher (FAA Technical Center), Frank Tullo, Ray Valeika, John Stelly Jr. (Continental Airlines); John Goglia, Joe Kania, and Dave Driscoll (USAirways); Joan Kuenzi (Northwest Air Lines), Bob Doll, Lou Mancini, Yesso Tekerian, Jeff Leeds, Lenny Page, and Jim Griffiths (United Airlines); Richard Kormaniski (Grey Owl Aviation), Art Columb (Air Transport Assn), Dave Cann, Jim Ballough, Vince Lepera and Al Zito (FAA FSDO19); Dean Bill Petek; Professors Naj Meshkati and Mike Barr (USC Institute of Safety and Systems Management); and Dean Terry Stroup (Santa Clara University School of Engineering). Michelle Robertson was an energetic and productive partner in much of this research.

And finally there are the others—the thousands of airline maintenance technicians, members and representatives of the IAM-AW, foremen,

Acknowledgments

managers, and staff support people—who are willing participants in the research and who have given so freely of their time and experience. It is to them that this book is dedicated. Despite all this help we, of course, remain entirely responsible for mistakes or inaccuracies that remain.

> Jim Taylor
> Palo Alto, California
>
> Tom Christensen
> South Bend, Indiana
>
> July, 1998

Contents

Foreword   xi
Introduction   xiii

Chapter 1
What Do We Mean When We Say "Communication"?   1

OVERVIEW: The Subject Is Workaday Communication, Not "Feelings"   1
Safety Depends on It   1
What You Need to Know Right Now   3
Here's the Logic   4
More Logic   7
And More   7

Chapter 2
Must AMTs Be Skilled Communicators?   11

OVERVIEW: Better Communication Skills Are Required Today   11
Traditional Selection Criteria   11
What Are AMTs Really Like on the Job?   12
What About the Lead Mechanics?   13
Why Do AMTs Make Mistakes?   15
Remember: Workplace Communication Is About the JOB   18

Chapter 3
Communication Mistakes in Maintenance Costs Lives   21

OVERVIEW: Maui and Eagle Lake Accidents Focus Attention on Maintenance Communication   21
Outline of the Accident   23

    The Company's Larger System   24
    Communication and Collaboration   27
    The Eagle Lake Accident   28
    Outline of the Accident   28
    Collaboration, Communication, and the Eagle Lake Accident   31
    Culture Is Hard to Change   33
    A Parting Thought   34

Chapter 4
A Short History of Communication and Human Factors Management in Aviation Maintenance   37

    OVERVIEW: Human Factors Management Is a Powerful New Science in Aviation   37
    What Human Factors Management Is All About   38
    The Evolution of Human Factors Management   40
    Culture Clash in Aviation Maintenance   45
    Cultural Change in Maintenance in the 1990s   47
    Beyond Human Factors Management   48

Chapter 5
Professionalism Now Includes More Communication   53

    OVERVIEW: A New Approach Is All But Inevitable   53
    Big Problem: Some Mechanics Now Just "Factory Hands"   54
    Specialization Reduces Professionalism   56
    Professionalism? In this Environment?   60
    Professionalism in Context   61
    A Parting Thought   63

Chapter 6
The Quality of Communication Determines the Quality of Decision Making   65

    OVERVIEW: Cost Pressures Threaten Communication Quality When It's Needed Most   65
    Outside Vendors Increase Risk of Costly Communication Errors   67
    Offshore Engine Overhaul Leads to Serious Injury and Major Property Damage   67
    Factory Worker's Propeller Repairs Lead to Serious Accident   69

The Deadly Effects of Mislabeled Cargo   71
Missing O-Rings Lead to In-Flight Shutdowns and Nearly Cause Disaster   74
A Parting Thought   78

Chapter 7
Up Your Professionalism   79

OVERVIEW: Today's Mechanics Must Rise to New Levels   79
Professional AMTs Need Better Communication Than Mere Grease Monkeys Do   81
Professionalism Defined and Delivered: The 4 Cs   83
New Classifications Mean More Communication Skills   84
More Professionalism or Less?   84
Some Parting Thoughts   87

Chapter 8
Mechanics Cannot Thrive on the Written Word Alone   89

OVERVIEW: Effective Communication Is Multimedia   89
Oral Communication Has Limitations   93
Written Communication Is As Important As Ever   94
The Accident at Dryden, Ontario   95
Encouraging Attempts to Improve Written Communication   100
A Parting Thought   103

Chapter 9
MRM Is Multiparty Cooperation, Open Communication, and Error Reduction   105

OVERVIEW: MRM—A Strong Model for Multiparty Communication, Cooperation, and Error Reduction   105
PHASE ONE: Participative Data Collection   108
PHASE TWO: Making Changes With and Without AMT Involvement   111
PHASE THREE: Broadening MRM's Scope from Paperwork Errors to All Quality   117
PHASE FOUR: Broadening MRM to Become a Culture   118
A Parting Thought: The Bases of Success   119

Chapter 10
Early Successes with Open Communication   123

OVERVIEW: Continental Airlines' MRM Communication Program Pays Off   123
CRM and Teamwork Training in Aviation   125
Continental Introduces MRM   126
The Importance of Purposeful Systems   138
A Parting Thought: Borrow from the Best   141

Chapter 11
AMT-Oriented Communication Training Comes of Age   143

OVERVIEW: The Need for Strategic Thinking and Management Follow-Through   143
Background   143
The Increasing Variety of Current MRM Programs   146
The New Age of MRM Programs   148
A Case Study in Bringing MRM to AMTs   150
Initial Training Evaluations Very High   152
Improved Attitudes   153
Performance Changes Related to MRM Training   157
Reported Changes from MRM Training   158
A Parting Thought: The Importance of Implementation   160

Chapter 12
Recommendation Number One: Change Your Mind   163

OVERVIEW: The Real Trouble with Maintenance Is Called Culture Lag   163
Achieving an Effective MRM Culture   175
A Final Thought   182

Appendix A
Principles of Cultural Redesign   185

Appendix B
Some New-Culture Programs Already in Place   189

Index   191

About the Authors   201

Foreword

If you care about aircraft safety, this is a book you must read—no matter what your connection to airline maintenance, no matter where you are in the world. This book is a primer about the leading-edge approach to maintenance operations, the partnership of manager, doer, and regulator—the art and science of Maintenance Resource Management (MRM).

As a former president of the Society of Automotive Engineers (SAE) and as the recently retired vice president of maintenance operations for USAirways, I have been twice blessed by having helped get this book commissioned and by having had the privilege of getting an airline started with MRM. Our own story is told in Chapter 9, warts and all. Despite our early successes with this new approach, USAirways is continuing to learn the ins and outs of MRM.

Taylor and Christensen have something important to say to all of us.

In clear and compelling language, the authors tell the inside story of maintenance communication errors that led to near disasters and also to airplane accidents that claimed hundreds of lives. Although these examples are drawn from North American experience, the lessons apply to maintenance operations everywhere.

The authors show how these sometimes simple workplace communication failures could have happened at any airline, anywhere, thanks to an industrywide maintenance culture that has evolved too slowly to keep pace with the changes in our industry. This book confirms that communication skills are the lifeblood of the aviation maintenance technician; they will help us address the ever-increasing complexities in our technology, government regulations, and the marketplace, in order to produce a continuously improving safety environment.

In the last three chapters of this book, the authors detail how the MRM programs at several leading carriers are reducing maintenance errors and improving the professional caliber of mechanics and managers. The authors give you the numbers, directly from the latest research on the effectiveness of these MRM initiatives. They "make no bones" about it—they show you how the deadly mistakes of the past are practically guaranteed to rise without these kinds of fundamental reforms in our old, familiar workplace communication practices.

The changes the authors are calling for are needed everywhere and at every airline.

Wherever the old ways remain in use, accidents are waiting to happen—accidents that do not have to happen. Such accidents are caused by maintenance errors that can be prevented.

Taylor and Christensen go one step further, reaching beyond the best practices beginning to take hold in our industry today. They also brief us on what lies ahead and what will be needed to bring us abreast of the high-performance work systems in the best high-tech industries around the world.

In one book, the authors cannot teach us everything we need to know. However, they can show us where to start—right where we are—and when to start—right now, today. If you did not already know why you must start moving, and keep moving, in this direction, they make it painfully obvious.

As for the pain, these authors have been around and know well that their prescription for change will not be easy medicine to swallow. In some ways, their prescriptions are a frontal assault on a way of working life that for us continues to feel tried and true despite having outlived its usefulness.

Some of our best stuff—the things that have worked so well in the past—are now becoming the biggest obstacles in our paths. Taylor and Christensen introduce us to the enemy—ourselves. To achieve what we want in the future, we must begin by changing our own minds right now. Rather not? Do it anyway, because lives are at stake.

It is an urgent message clearly stated. This message must be said and heard, loudly and clearly, throughout the industry.

A good place to start, or to pick up where you left off, is right here. Just turn the page.

B.R. Aubin
Senior Vice President, Technical Operations,
Air Canada (retired)
Senior Vice President, Maintenance Operations,
USAir (retired)

Introduction

The Critical Nature of Communication

In 1903, Orville and Wilbur Wright completed the first successful powered flight. Although their preparations seemed complex at the time, by our standards they were very simple. When the brothers designed, flew, *and* repaired their own airplane, they didn't worry much about communication breakdowns, because they were involved in every operation. Over the next 95 years, as advances in technology changed every aspect of aviation, little attention was paid to an important element of safety—communication in the workplace.

In 1988, the industry got a wake-up call from a fatal accident involving a Boeing 737 near Maui, Hawaii. One crew member died, and a number of passengers were injured when the skin on the aircraft peeled back in flight. The National Transportation Safety Board (NTSB) attributed the accident to the management of the entire maintenance system, rather than to individual error [Ref. 1]. Although maintenance issues had come up before, this accident marked the first time the NTSB laid the blame squarely at management's door. After the NTSB's report on the Maui accident, it began to dawn on the aviation community that communication is an industrywide problem. Even so, communication and coordination issues have only recently been considered as important as technological advances.

By definition, management is responsible for the communication and coordination systems of their organization. However, after airline deregulation in 1978, management's focus shifted to mergers and cost reduction issues. Employee needs and organizational structures were of secondary importance. Competition pressured airlines to pursue productivity improvements aggressively, to minimize "nonessential" maintenance activity, and to reduce costs, even as they expanded operations. As technology and economics governed airline operations from 1978 into the mid-1990s, aviation accidents continued at a nearly constant rate.

In 1996, aviation accidents worldwide accounted for over 1300 passenger fatalities, nearly twice the annual average of the previous five years [Ref. 2]. Today, it is estimated that air traffic will double in the next decade. So, just maintaining the current annual number of accidents requires that today's accident rate (1.5 accidents per million flights) be cut in half. With the enviable historical safety record of commercial aviation, such a reduction is a daunting challenge. However, accident studies indicate that improvements in communication can reduce this accident rate.

Because human interactions are so vital to safe aircraft operation and these practices need to be improved, we analyze numerous actual events for shortcomings in communication and coordination activities. Additionally, we offer recommendations aimed at correcting these problems.

Why haven't these problems been solved? We could argue that getting people to communicate better, to consider other employees as part of the work system, and to pay attention to the hazards lurking in complex work environments should not be as difficult as designing and developing aircraft. But, although the solutions seem logical, the transformations required aren't easy for management to swallow. The desired holistic approach requires that management and organizational cultures change, sometimes dramatically. Unfortunately, "change" encounters resistance from employees at all levels. Our comfort zones must be expanded; to do this, we must break down social distances and involve all employees. To many aviation maintenance personnel, such a transformation is equivalent to changing their identity.

A desirable trait in the past, individualism can be a problem in the current safety environment. Those involved in aviation safety must learn to work as teams, and they must reform their linear communication style. This is an especially difficult barrier for maintenance employees. With their engineering focus, maintenance managers and technicians possess highly technical skills, but sometimes lack the communication skills necessary to ensure safety in today's complex operations.

What is needed is a better balance of technical skills and social skills. Workplace communication must be improved if the job is to be done right.

Supervisors, leads, and staff must continually strive for excellence in communication. Furthermore, new programs must be designed to accommodate worker needs and play to their strengths.

Fortunately, airlines realize that a fix exists for maintenance communication problems. They are addressing this problem through more effective sharing of information among all employees. This "bottom-up" technique is called Maintenance Resource Management (MRM). Addressing the growing crisis in communication, MRM is evolving quickly. During the short history of its implementation in North America, there are already several success stories. This book outlines recent triumphs and failures of the successful migration from antiquated communication to ground-breaking, employee-motivated teamwork. Practical recommendations for further improving workplace communication are also included.

Following the MRM programs enables maintenance employees to take on new responsibilities, *and* it makes their jobs easier, not harder. Resistance to change is fading away as employees see tangible benefits. Companies involved in developing new programs are cooperating with their competitors to learn how these programs can benefit the entire industry. Indeed, intra-industry cooperation inspired this book.

Our goal is to help aviation maintenance employees meet their ongoing and unchanging mission: Keep airplanes safely in the air. We know that better communication saves lives, improves on-time performance, and enhances cost management programs. The days of the Wright brothers and the cult of the individual are fading; only fully cooperative teams can accomplish this crucial mission. We need effective, strong workplace communication at every level.

John J. Goglia, Member
National Transportation Safety Board

References

[1] National Transportation Safety Board (NTSB), *Aircraft Accident Report; Aloha Airlines, Flight 243, Boeing 737-200, N73771, Near Maui, Hawaii, April 28, 1988.* NTSB/AAR-89/03, Washington, D.C., 1989.

[2] "How safe is your airline?" and "Fasten your safety belts." *The Economist* (January 11, 1997): 13–14, 55–57.

Chapter 1

What Do We Mean When We Say "Communication"?

OVERVIEW: The Subject Is Workaday Communication, Not "Feelings"

When we're talking about communication in this book, we're talking about work-related communication. We're talking about the information exchanges that make a difference in the safety and efficiency of a mechanic's work. We're talking about the ways communication *about* the work affects the quality, quantity, and cost *of* the work. We're not talking about "feelings."

We're talking about information, coordination, and teamwork—on the job and about the job—on the flight line, in the overhaul hangar, and in the repair shops too. This kind of regular, workaday communication is one of the most important and least understood of the so-called "human factors" affecting output.

Safety Depends on It

The need for accurate and timely information, for coordination, and for teamwork, is increasing in aircraft maintenance, just as it is in many other occupations today. Better communication is needed now because the work itself is becoming ever more complex [Ref. 1 pp. 83–90; 2 pp. 35–37].

Because of this complexity, aviation is increasingly unforgiving of human error, not just pilot error. Lives have been lost because of maintenance errors—errors made because of inadequate communication. In later chapters we'll show you how inadequate shift turnovers and insufficient communication between management and the technicians have created maintenance nightmares that have led to fatal accidents.

Greater complexity requires greater communication. Time compression multiplies complexity. Instead of having ample time to think and plan, many modern workers are required to make increasingly complex decisions more quickly. The impact of these decisions is often greater than in the past. Fast decisions about complicated situations affects more people, too, so the stakes are rising.

Complexity, time pressures, high stakes—these have all existed in maintenance for a long time. What's new is their sharp and radical increase. The new levels of complexity, pressure, and high stakes are fast outstripping the old communication system's ability to cope. The airlines must design and implement new systems soon, for one very good reason—because they'll have to.

The complexity of maintenance work has grown beyond the grasp of the individual working alone, so more and more often no one person can comprehend the whole job or coordinate all of its parts alone. The old communication system relies heavily on the decisions of the individual general mechanic working alone. This system has worked pretty well for a long time. But that time is nearly over.

The accidents we discuss in the next chapters will lay it out for you. The cause of many accidents is found in the larger communication *system* and *not* in the mechanics themselves alone. A mechanic's individual decisions, based on limited information and little time, simply cannot meet the challenge of today's complexity. As we'll show you later, a poorly informed decision by just one mechanic can bring on disastrous consequences. And no mechanic—not one—can reasonably be expected to have all the pertinent information about everything they might encounter on the job.

It is neither safe, nor even possible, for maintenance managers or foremen to coordinate truly complex work all on their own. It can't be done any more. To know everything is simply too much for any one person. Today's job is even harder because important information is scattered. Key technical information, along with everyone's differing experiences with it, is not always readily available so that employees can make well-informed decisions.

Greater complexity is thus driving the need for a more sophisticated approach to workplace communication—a more wholistic approach. Fortunately, such an approach exists. It is more sophisticated and more effective, but it adds surprisingly little to the organization's existing complexity. In fact, it often reduces complexity. Once understood, designed, and put into place, the new communication systems are models of simplicity.

What You Need to Know Right Now

1. Regardless of the media we use, communication practices in any organization are intended to provide needed information so the purpose of the organization can be met [Ref. 3]. Some communication systems do this better than others.
2. Workplace communication is not merely the transfer of data from one point to another. When communication occurs, it creates a live circuit, a dialog that creates shared understanding between the people at both ends of the wire (or at every connection on a "party line") [Ref. 4]. Effective communication creates common ownership of the task at hand and shared objectives that enable people to control the work process even if they later have to make an important decision alone. There are fewer false assumptions, fewer overlooked details, fewer mistakes [Ref. 5 pp. 12–17; 6 pp. 83–86].
3. To ensure good decision making and common understanding, basic workplace communication is best grounded on informal team meetings that include everyone involved [Ref. 7; 8 p. 248]. Don't panic; it pays far more than it costs.

Short meetings with the whole team are more efficient *and* more effective than when the lead or supervisor meets with each person individually, or when individual mechanics meet one-on-one with other individual mechanics to pass on turnover information. That's the old way—the domino effect: Knock over the first domino and all the rest are expected to fall in predictable order. Sure it looks efficient on paper, or in the abstract, but people are not dominoes. Even in the most rigid command-and-control hierarchies, even a slight misunderstanding at any level can rapidly spread increasingly erroneous messages in every direction [Ref. 4; 9 p. 8].

A short meeting of an entire team gets far superior results—and often takes less time—than relying on one-to-one communication strategies alone. The information can then be pooled by all of those present. Assumptions can be checked; questions asked and answered; suggestions made; and even the physical state, body language, and eye contact of the participants can be scanned for what information they might reveal that could be vital to the safety of the maintenance people and the plane in front of them.

The domino effect often produces messages at one end that were not intended at the other. Sometimes the results are as laughable as the party game "telephone," except that airplane safety is never a laughing matter. One-to-one communication up and down the line almost guarantees that very unfunny things will happen unexpectedly. Serious give-and-take meetings of the entire team can reduce misunderstandings, poor decisions, and dangerous errors; such meetings improve not only the knowledge base, the decision skills, but also the attentiveness of the team members [Ref. 10].

In times past, all of the existing complexity could be contained in the head of an intelligent mechanic, working alone. But the vastly expanded complexity of maintenance work today requires that *all* of the information be shared among *all* of the people assigned to *all* of the repairs on the entire aircraft. This information is needed to ensure that the job is tackled as a whole—as an interconnected, interactive, whole system [Ref. 7].

Better decisions, better coordination and fewer errors are the payoffs. Shown time and again in many other industries [Ref. 11], this is a new point of departure for aircraft maintenance, and it requires a shift in thinking that needs further explanation.

Here's the Logic

This new approach departs from the one-to-one communication pattern favored by aircraft maintenance in the past. Instead, modern communication practice favors a "whole-system" view of workplace communication. It recognizes that the entire set of technical and human factors involved must be managed as an interactive whole.

Airplanes, manuals, tools, training, management, and the Aviation Maintenance Technicians (AMTs) themselves are all *parts* of the maintenance system. Taken as a whole, they *are* the system. It is the interaction among these parts that get the job done, right or wrong. It has *always* been an error to think otherwise. From this point of view, the system is understood as much more than a simple string of causes and effects; it is a complex of intertwined, interacting, even mutually determined components. It's not a mechanism; it's an organism.

Unfortunately, few airline companies manage to put this insight into action. Most seem addicted to an outmoded approach that follows the mostly top-down, shift-to-shift, domino effect mentioned earlier. This is straight-line, mechanical thinking. It seems logical and efficient. And up to a point it works.

But this approach has a hidden, and sometimes fatal, flaw.

It's linear. By sticking to the straight and narrow path, the domino approach ignores the multiple, mutual interactions that drive the system as a whole. It ignores whatever seems irrelevant, not strictly to the point. What seems irrelevant, however, may not be irrelevant at all. That is why it's not uncommon for the old standard communication to create all manner of misunderstandings, errors, and confusion. Too much is left out of the equation.

Remember this old story: For want of a nail (irrelevant!), a shoe was lost. For want of a shoe, the horse was lost. For want of the horse, the rider was lost. For want of the rider, the battle was lost. Remember that?

For systems thinkers, everything is relevant; every outcome is "sensitive to initial conditions" [Ref. 4; 9].

It's easy to see why domino pushers have a hard time finding the causes of certain mistakes. They don't look at the system as a whole, nor even at their own part in how it functions overall. But the times are changing.

To summarize: Old-time practices in workplace communication are increasingly ineffective. They do not deliver the best results. They are themselves an ever-larger part of the problem. They expect too much of management, and they expect too little of their main partners in the system, the AMTs.

In the following chapters, we'll show you just how much the old system is like a worn-down creaky machine, a steam engine that needs more than oiling and a few new parts. It needs to be replaced. The system of the future is more like a pride of lions—it's alive and self-renewing.

The new approach puts a whole, systems view of workplace communication into action. System members are brought together often. The system is understood to include everyone involved in the work—usually meaning everyone, on every shift, who will work on any part of the aircraft, in concert with the associated management and support staff. This larger team defines the work required and plans for its accomplishment together.

That's the skeleton of an effective maintenance communication system that works. It handles complexity better because it harnesses the information and knowledge of the whole organization at once. It prevents misunderstandings, puts everyone in the know, and creates real teamwork and outstanding results.

That's why whole-system communication is so much more powerful than any domino approach [Ref. 12; 13]. It gets the job done better—more efficiently and with less chance of error.

AMTs usually work at a distance from one another and from their leads and supervisors. They are continually faced with time constraints and with work changes. But the general rule is the same whether it's a dispersed shift crew or a small work group or even the whole company that's involved: "Get everyone into the same room," into the same discussion, into the same current of information. Get them involved in communicating with each other about the purpose of the work and the best ways to get it done [Ref. 3; 8].

This approach does not aim to make everybody feel good. This approach makes sure that the work itself is good. Interestingly enough, high-performance teams generally feel good about their work and about themselves. But their emphasis is on the work they do together and only indirectly on how well they get along together. In itself, achieving real results improves human relations, but improving human relations alone does not achieve significant or lasting results.

So, focus on the work itself (the whole job) and on the work team meeting (the whole team). Oh sure, there are lots of other things to think about, but you'll never go wrong keeping your eye on these two halves of your socio-technical system. Effective work is the product of a system that is both "socio" and "technical." To overbalance one way or the other is to fall down on the job.

More Logic

The logic supporting the work team meeting approach as the first line of action for improving communication and performance in maintenance is very simple.

First, every individual involved in the work may have unique, and even vital, information important to the work of the group as a whole.

Second, to make the best use of this information, all the people involved must be able to share what they know with the others.

Therefore, put the information and the people in the same place at the same time. Tackle the work as a whole team. It's done in other industries, so why not aircraft maintenance? Maintenance loves logic. And whole-team meetings and whole-system communication are logical.

Sometimes a 10- to 20-minute stand-up meeting in the hangar might do the job. "All the people" might mean the people on one plane, or the people on one shift, possibly overlapping to create a meeting of everyone across two shifts. Sometimes it means you need to get all 300 people (or whatever) into a hangar someplace and after that they can form a "virtual" team communicating by document, e-mail, telephone, or fax. There is more than one way to get everyone into the "same" information stream. Each has its place. None is all that hard to do.

And More

This is the systems approach to workplace communication in its simplest form. Everyone's information is more complete, more accurate, and more useful for decision making. Better decisions produce better work—better quality, better productivity, and even better job satisfaction.

This kind of communication transforms the raw materials of the job into gold. Honest. A whole-system communication approach, including group decision making, can and does raise the market value of every part of the system: the technology, the product, and the people involved. Every part becomes more valuable.

It takes new skills and hard work to do it right, but team-based communication and decision making are the most direct and powerful ways to improve business results and employee achievement. This has been demonstrated many times over, in industries as varied as autos, steel, food, natural resources, health care, environment, and government service, to name a few [Ref. 11]. These industries are reporting 15–100% improvements in such key measures as quality, quantity, and cost [Ref. 10]. It works for them. It deserves a good try in aircraft maintenance.

To introduce a team-based system in maintenance, AMTs and their managers will have to learn the technology of modern workplace communication, including shared responsibility for coordination and for true teamwork (versus merely *feeling* like a team or merely working in the company of others). For some people, that is scary. And for good reason.

Team decisions are not necessarily the only right decisions. It can be frustrating and threatening to have to yield to a consensus instead of garnering support for our own ideas. It often feels better to make our decisions in private rather than exposing our thinking to our colleagues' appraisal.

Sometimes team members find themselves in conflict; this is uncomfortable until the conflict is resolved. But seeking consensus, surfacing and sorting through conflicts and differing points of view—these are the very things that true team players come to welcome. These are often the sources of new knowledge, better understanding, and better decision making—better teamwork too.

AMTs who take their communication and decision-making responsibilities seriously are clearly likely to better understand both the details of the work and the larger picture. Working together in this way, they often generate better coordination, less waste, fewer errors, and smarter risk-management judgments than the individuals involved could have done on their own.

What's more, team-style communication creates better accepted and better implemented decisions. At the very least, everyone involved understands the context of the work and is aware of the actions being taken by others. Individual decision making is better informed. And people in nearby areas are better tied-in.

Face-to-face and "virtual" team meetings about the work at hand are the fundamental elements in any modern maintenance communication system.

In addition to using this approach for the regular work, the same pattern can be effectively applied to special projects such as improving log books, technical training programs, and engineering orders. Every time a team-system approach is used, it improves the organization's ability to make even better use of teams in the future. Team-based communication is a self-regulating, continuous-learning system [Ref. 13]. It may also be the best training for the dollar your company will ever buy.

Frequent group meetings to review the work at hand are increasingly beneficial as team members gain skill in this kind of communication and decision making. But such meetings are not merely beneficial, they are necessary. Unfortunately, most aircraft maintenance departments do not think highly of this style of communication as a useful and necessary business discipline. Nor do the majority of AMTs prefer to communicate directly in a group. So it is not simply a matter of improving skills; the industry's culture must change.

Culture is another term for communication system, with the extra punch of emotional investment and deep-seated habits, values, and beliefs. A new approach to communication involves a powerful cultural change that cannot be avoided in aircraft maintenance today. Lives depend on it. If it's any comfort, cultures throughout history have had to change to adapt to new circumstances in the world around them. For maintenance, the time to change is now. Right now.

References

[1] Fox, Matthew. *The Reinvention of Work: A New Vision of Livelihood for Our Time.* San Francisco: Harper, 1994.

[2] Cotter, John J. *The 20% Solution: Using Rapid Redesign to Create Tomorrow's Organizations Today.* New York: Wiley & Sons, 1995.

[3] Taylor, J.C., and D.F. Felten. *Performance by Design: Sociotechnical Systems in North America.* Englewood Cliffs, NJ: Prentice Hall, 1993.

[4] Toffler, Alvin. *The Adaptive Corporation.* New York: McGraw Hill, 1985.

[5] Peters, Tom. *Thriving on Chaos.* New York: Knopf, 1987.

[6] Naisbitt, John, and Patricia Aburdene. *Reinventing the Corporation: Transforming Your Job and Your Company for the New Information Society.* New York: Warner, 1985.

[7] Taylor, J.C.; P.W. Gustavson, and W.S. Carter. "Socio-technical Design and the Adaptation of Computer-Aided Design (CAD) Processes," in D.D. Davis (ed.), *Implementing Advanced Manufacturing Technology.* San Francisco: Jossey-Bass, 1986.

[8] Weisbord, M.R. *Productive Workplaces: Organizing and Managing for Dignity, Meaning and Community.* San Francisco: Jossey-Bass, 1987.

[9] Gleick, James. *Chaos: Making a New Science.* New York: Viking, 1987.

[10] Christensen, T.D. "A High-Involvement Redesign," *Quality Progress Journal,* May 1993.

[11] Lawler, E.E., III. *High-Involvement Management: Participative Strategies for Improving Organizational Performance.* San Francisco: Jossey-Bass, 1986.

[12] Maurino, D.E., J. Reason, N. Johnston, and R.B. Lee. *Beyond Aviation Human Factors.* Aldershot, Hampshire: Ashgate, Avebury, 1995.

[13] Senge, P.M. *The Fifth Discipline: The Art & Practice of the Learning Organization.* New York: Doubleday, 1990.

Chapter 2

Must AMTs Be Skilled Communicators?

OVERVIEW: Better Communication Skills Are Required Today

Aircraft Maintenance Technicians (AMTs) are required to have high-level technical skills, but little or no consideration is given to their communication skills. This chapter will present the results of several studies. They show how little attention has been given to communication, how important communication is, how little communication skills have been developed, and how both AMTs and maintenance management see this lack.

Traditional Selection Criteria

A 1984 U.S. government careers pamphlet featuring aviation maintenance [Ref. 1] describes the work, its benefits, and the individual requirements for entering and advancing in this occupation. Table 2.1 lists these entry-level requirements.

According to the pamphlet, advancement beyond the Aircraft Mechanic Certificate includes jobs as lead mechanics, crew chiefs, inspectors, or shop foremen. Promotion to these higher-grade jobs is usually attained as a result of technical skill plus company service and seniority. Further advancement is restricted to a "few mechanics with advanced ratings and administrative ability" [Ref. 1].

In this view, the mechanic's job is defined by its technical requirements only, without a word about the important human elements that are involved such as face-to-face communication, coordination, and teamwork—likewise for the jobs of the leads and chiefs. Communication skills are thus not

TABLE 2.1 ENTRY-LEVEL REQUIREMENTS FOR AN AVIATION MAINTENANCE CAREER (1984)

- Understanding the physical principles involved in the operation of the aircraft and its systems
- An understanding of English
- A willingness to continue technical education
- An above-average mechanical ability
- A desire to work with one's hands
- An interest in aviation
- An appreciation of the importance of doing a job carefully and thoroughly

Source: FAA Careers Pamphlet [Ref. 1]

mentioned as a requirement for mechanics, or for leads, inspectors, foremen, or any jobs higher up the ladder. We challenge the accuracy and usefulness of Table 2.1's outdated list by offering the following alternative list of AMT characteristics and views, collected from a variety of sources.

What Are AMTs Really Like on the Job?

AMT job descriptions are one thing. The people who actually meet these requirements tend to share certain other characteristics as well. General observations of what AMTs are actually like on the job have been circulating for many years. One thing common to these lists is a distrust of words together with the absence of communication skills. In fact, a certain resistance to communication appears to be part of the AMT profile. One widely accepted list of AMT traits is shown in Table 2.2.

The list in Table 2.2 says a lot about AMTs and their general attitudes toward communication. Basically, AMTs are strong-and-silent types, loners who seldom communicate at all, much less in group discussions. And AMTs have themselves come to believe that they cannot communicate very well for the simple reason that their work system does not expect them to be able to communicate very well.

Many studies (see Chapters 9–11) have found that once the impediments to open communication are removed, AMTs are as communicative as anybody else. The fact is that in their professional life they are normally in a

TABLE 2.2 WHAT CHARACTERIZES THE AMT?

- Commitment to excellence
- Willingness to put in effort and hours
- Integrity
- Distrust of words
- Dependability
- The tendency to be a loner
- Modesty (no desire to be in the spotlight)
- Doesn't like to ask for help
- Tends to be self-sufficient
- Doesn't share thoughts too frequently

Source: Richardson Management Associates [Ref. 2]

"one-down" position. Management, pilots, and engineers all look down on them—or at least that's the way many AMTs see it. Working as equals with these people brings out the best in the mechanics and in the system as a whole.

What About the Lead Mechanics?

A recent study of lead mechanic selection and training in a large U.S. airline [Ref. 3] confirms the skills and skill gaps predicted by the selection guidelines in Table 2.1, on the one hand, and the AMT characteristics presented in Table 2.2, on the other. According to the study, lead mechanics were originally nominated and selected by a majority vote of their peers; this was later replaced with peer nomination and management selection.

The study examined the training requirements for newly selected lead mechanics. It also surveyed 62 lead mechanics (43% of all leads in the targeted departments) about the most challenging aspects of their jobs, those aspects for which they were least prepared, and their conclusions about the most important skills for their jobs. The results, shown in Table 2.3, indicate that interpersonal workplace communication and other communication issues are clearly the most challenging skills and also those for which they feel least prepared. Add that to the expectation of little communication, and you have a pretty good definition of the problem.

TABLE 2.3 HOW LEAD MECHANICS' VIEW THEIR JOBS AND TRAINING NEEDS (*n* = 62)

What is the most challenging part of your job?	Percent
Human relations/dealing with people	36
Meeting "ready times"	19
Motivating mechanics	11
Coordinating different departments	8
Workload (# of aircraft)	7
Dealing with the corporate "system"	7

What aspects of your job do you feel you were least prepared for?	Percent
Human relations/dealing with people	37
Paperwork	19
Computers	18
Lack of training	8
Many skills to know (avionics, etc.)	7

What are the most important skills you need to be effective in your current position?	Percent
People skills	48
Communication skills	44
Technical skills	27
Computer skills	16
Listening skills	10
Organizational skills	8
Decision-making skills	8
Management/supervisory skills	8
Leadership skills	7

Source: Predmore and Werner [Ref. 3]

Why Do AMTs Make Mistakes?

Several studies shed light on why mistakes occur. It is not unusual, however, for mechanics and managers to differ on what communication and job performance issues are involved. This is in itself a sign of serious communication problems.

One study cited by Boeing [Ref. 4] lists the most important causes of errors as identified by 150 mechanics. Table 2.4 shows that "poor communication" is an important cause of mistakes. In addition, at least three other causes in Table 2.4 are clearly communication breakdowns: "failure to understand [communicated] instructions," "lack of available [not communicated] instructions," and "pressures [communicated] from management to defer work." AMTs understand that communication, or its lack, is directly related to errors in the work.

Another study, conducted in one large airline during 1992–3, examined the sources of AMT errors in documentation and paperwork [Ref. 5]. In this study, 160 foremen, lead mechanics, and AMTs brainstormed a list of errors

TABLE 2.4 AMTs' VIEW OF WHY THEY MAKE MISTAKES (150 AMTS)

- ◆ Boredom
- ◆ Failure to understand instructions
- ◆ Lack of available instructions
- ◆ Rushed
- ◆ Pressures from management to defer work
- ◆ Fatigue
- ◆ Distractions at a critical time
- ◆ Shift change
- ◆ Poor communication
- ◆ Use of incorrect parts
- ◆ Poor lighting
- ◆ Failure to secure fasteners
- ◆ Unauthorized maintenance

Source: The Boeing Company [Ref. 4]

and their causes in 30 group interviews (20 for AMTs, 10 for leads and foremen) in seven line maintenance stations. Only the top 20% of all causes listed were judged "most important" sources of the errors. Table 2.5 lists the most important causes named in this study. Despite the difference in focus (paperwork errors versus all maintenance mistakes), this list is similar to that found in Table 2.4: trouble understanding instructions, pressures and distractions from management and other departments, and poor communication with upper management and with other maintenance sections.

It is clear from these studies that AMTs, leads, and foremen recognize that they are operating in a less-than-perfect communication culture. As for their own roles in this, they acknowledge that they often act mainly as individual contributors, and not as part of a truly communicative culture. This is an interesting self-contradiction: They like working alone, yet they complain that their biggest problems stem from lack of communication with others. Are they ready to make the change to a new team communication culture? No one has asked.

AMTs realize that many of their most serious errors come from poor communication practices throughout the entire maintenance system. The system thus has to change to cut that list of errors down to size. Since the system includes numerous interacting parts, including many points of view, maintenance management and AMTs must reconcile their differing opinions about the causes of AMT errors.

While the AMTs see things one way, Boeing reports that airline management sees a different picture. Specifically, managers at several airlines told Boeing that errors are caused by:

1. AMTs having a lack of time
2. AMTs acting unprofessionally
3. AMTs' lack of training
4. Inexperienced AMTs [Ref. 4]

Many aviation maintenance executives hold the same views as the managers interviewed by Boeing. They simply don't see communication as a problem, and certainly not their problem. It's not even on the list. Instead, AMTs themselves are at the heart of the list. AMTs put *management* communication shortcomings high on their list of causes of errors. Management puts *AMTs* at the top of their list. Clearly, they need to sit down together and try some of that whole-team communication.

TABLE 2.5 AMTs' VIEWS OF WHY THEY MAKE PAPERWORK MISTAKES (160 AMTs, LEADS, AND FOREMEN)

◆ *Communication about technical information.* Upper management is not interested in communicating to lower management or hourly personnel. Queries made of upper management are reported to result in few explanations or reasons for changes, discrepancies, or inconsistencies in the General Maintenance Manual (GMM). Respondents believe problems in the GMM cause errors in paperwork.

◆ *Maintenance system practices regarding information.* There is dissatisfaction with repair and signoff advice by Maintenance Control, and with the maintenance planning department issuing incorrect phase check paperwork. Paperwork is reported to be often lost or misplaced in the maintenance planning or aircraft records departments.

◆ *Merger-related information/communication issues.* Respondents felt that management, by their actions, did not consider using documentation procedures (some of which were felt to be clearly superior) from other airlines it acquired.

◆ *Information about transit checks during the day.* There is not enough time to accomplish transit checks and the associated paperwork during the operating day.

◆ *Log book design and use.* The log book modifications have not involved mechanics, and there was no subsequent training on those changes. Log book design was reported to cause certain signoff and data errors.

◆ *Complex and/or redundant engineering information.* Errors are caused by overly complex, unnecessary, and/or redundant engineering information, contained in Engineering Orders and Airworthiness Directives (ADs). The engineering, programs, and technical publications departments don't ask for input from AMTs before releasing documents.

◆ *General paperwork design and practice.* There is not enough time for untrained AMTs to complete the company's complicated paperwork without making errors.

◆ *Clarity of manuals and their use.* Policy manuals are not clearly written, hard to access, and difficult to use—paperwork errors result as a consequence.

◆ *Training in paperwork.* Neither initial paperwork training nor recurrent training is sufficient for AMTs to always complete paperwork in a timely or correct manner.

◆ *The type and condition of maintenance information technology.* Microfilm machines are antiquated and information provided (especially IPC) is distorted and/or unclear. The computer system is not user-friendly and not as integrated as it should be.

Source: Taylor [Ref. 5]

Everybody has an important stake in preventing errors, but better results will require a clear understanding of the whole picture and a keen sense of shared responsibility. Neither of these worthy goals can be achieved without improving workplace communication up and down the organization.

Remember: Workplace Communication Is About the JOB

Management and AMTs do not really have a problem with each other. Instead, they both have an incomplete understanding of the causes of maintenance errors, and they share the problems associated with their old-fashioned communication system. It's not a people problem; it's a problem with their ability to communicate technical information.

The human relations here don't need to be fixed. But the ineffective way of communicating straightforward, work-related information does. Focus not on people as the problem—management people, technical people, or even both. Focus on the job, on communicating work-related information across the board in order to improve results on the job.

We feel the need to repeat ourselves on this point. The issue is *not* one of human relations. The issue is the work itself. If the work-related communication system is improved, human relations will improve on their own. If human relations ever do appear to be the main problem, it's the result of poor communication about the work.

Many AMTs and their managers still do not see "communication" in that way. Some still call any communication-improvement program "charm school" or "Hug-A-Tree 101."

Forgive them, for they know not what they say.

AMTs work alone most of the time. The only team-based communication they experience at work may be discussions of someone's recent mistake and discipline. What do most of them know about communicating effectively with others, especially in groups? Few have had positive experiences with team-based, job-centered communication. They won't know what we mean by "communication" until they have to do it.

They'll be doing it pretty soon.

As we'll show in later chapters, several U.S.-based airlines have already begun to change their communication cultures. One company has trained all 2200 of its maintenance managers and maintenance staff professionals in better communication techniques. After three years, the results of open communication on safety and reliable performance are positive. Another company is training all 9000 of its AMTs and is over halfway there. A third airline is expanding on the approach of the first two and training all its maintenance personnel in safety awareness and communication skills.

Communication isn't just for other industries anymore. It has arrived in commercial aviation maintenance and it is proving a success.

References

[1] FAA *Aviation Careers Series: Aviation Maintenance,* GA300-123-84 (1984).

[2] Richardson Management Associates, Ltd. (2054 Sherbrooke St. West, Montreal, Quebec H3H 1G5, Canada), cited in K.D. Forth, "Squeaky wheel gets grease—and training." *Aviation Equipment Maintenance* (August 1993): 56.

[3] Predmore, S.C., and T. Werner. "Maintenance human factors and error control." *Proceedings of the 11th FAA Meeting on Human Factors Issues in Aircraft Maintenance and Inspection.* DOT/FAA/AM-, Office of Aviation Medicine, Washington, D.C. (1997): 79–89.

[4] Boeing Airplane Safety Engineering. *Model Maintenance Safety Program* (Paper No. UM2165). The Boeing Commercial Airplane Group, Box 3707 Seattle, WA 98124-2207 (1996).

[5] Taylor, J.C. *Maintenance Resource Management (MRM) in Commercial Aviation: Reducing Errors in Aircraft Maintenance Documentation* (Final Report of 1993–1994 Project Work). Socio-technical Design Consultants, Box 163, Pacific Palisades, CA 90272 (December 1994).

Chapter 3

Communication Mistakes in Maintenance Costs Lives

OVERVIEW: Maui and Eagle Lake Accidents Focus Attention on Maintenance Communication

The most challenging communication problems extend beyond the boundaries of a single department or company. In 1988, the aviation maintenance community had no choice but to consider how its communication system worked as a whole. And again in 1991, before much action had been taken to address that system, a second event occurred to further its examination.

The initial cause of this serious introspection was a series of communication errors that led the Aloha Airlines' maintenance department to overlook a number of small cracks in the fuselage of a 737 aircraft. The cracks suddenly enlarged in flight, which caused the hull to rupture and resulted in the death of one flight attendant and many injuries to other crew and passengers. The disaster could have been worse except for the skillful and prompt action by the flight crew, who brought the severely damaged aircraft down to a safe, but unscheduled, landing at the Maui airport. The second event, some three years later, involved the fatal crash of a Continental Express commuter aircraft at Eagle Lake, Texas. A chain of communication errors led to the failure to reinstall some 47 screws on the top surface of the plane's tail section, which caused its disintegration during a subsequent flight, killing all aboard. What might have been learned from the Maui accident to avoid the Eagle Lake crash? This chapter reviews the two cases and draws some parallels between them.

Although the Maui accident appeared at first to be an anomaly, employees and managers at other airlines could see that their practices could also cause similar results, if not in fuselage cracks then in other critical areas. The problem was with cracks this time, but it could be something else next time, all the result of similar industrywide communication gaps.

Indeed, that's how the National Transportation Safety Board (NTSB) saw the Maui case. Rather than merely blaming the inspectors for not locating the fatigue damage and the disbonding responsible for the rupture, the NTSB placed probable cause on the company's *overall* maintenance program—on its total package of systems, programs, and decision-making apparatus [Ref. 1]. This assessment gave pause to maintenance managers throughout the airline industry.

Only once before had an entire maintenance program been indicted by the NTSB [Ref. 2 p. 50]. In the Maui accident report, however, the NTSB went even further. It found that not only had the company management failed to recognize the communication and other human performance problems, there were also indications that "a similar lack of critical attention to inspection and detection was an *industry-wide deficiency*—(emphasis added) [Ref. 1 p. 72].

This revelation ran like a shock wave through the industry.

How different was that company's management from the other airlines? How different were their planes, their routes, their financial health? While such differences may have played a role in the accident, there were also plenty of similarities with other airlines, including the fact that many of Aloha's maintenance management practices were exemplary.

The following outline of the accident is offered as a kind of checklist for other airlines. The events are instructive in their own right, but perhaps especially so if you imagine your company's name in place of Aloha's. Could this just as easily have been your company?

Outline of the Accident

On April 28, 1988, Aloha Airlines Flight 243, a 737-200 en route from Hilo to Honolulu was flying at its level altitude of 24,000 feet when the fuselage ruptured just behind the left forward entry door. A section of crown skin ripped over the top to the other side and lifted from the belt line below the windows (at stringer 171).

Small cracks in the middle line of rivets down the belt line had become connected over a distance of 18 feet, allowing a major portion of the crown skin and structure to separate completely from the aircraft.

The flight attendant who died was immediately swept out with the initial rupture. Of the 94 people remaining on board, 65 were injured, 8 of those severely.

At the time of the accident, the aircraft had accumulated 35,496 flight hours and 89,680 flight cycles (take-offs and landings), which was the second highest number of cycles in the worldwide B-737 fleet. That alone should have flashed warning signals.

As was subsequently made clear in the NTSB report, the company had been lax in the supervision of its fleet maintenance program. Even a safety board member who dissented from the NTSB assertion of probable cause specifically cited the carrier's "failure to supervise its maintenance force properly" as contributing to the accident.

The NTSB report also included previous evaluations made by an FAA National Aviation Safety Inspection Program (NASIP) audit and one by a Boeing Customer Support Team as well. These two independent audits identified the company's need to develop a better management organization to establish and maintain controls and to provide more technical training and support for its maintenance workforce.

The Boeing audit further called for management development and leadership training for supervisors and greater participation by AMTs in safety training programs. Boeing also included a separate section on communication in which it recommended that the maintenance organization "work in harmony and in unison through the cultivating of proper communicative aptitude and attitude among the staff"—in other words, the company needed to encourage teamwork.

The NTSB report, including the Boeing and NASIP evaluations, contains a number of observations on the maintenance organization and its management policies during the period of the accident. These are listed in Table 3.1.

Table 3.1 shows a business-focused company with a human face. The company's relations with its individual AMTs was apparently very good. The record shows both harmonious labor relations and generous benefits for high-service employees. Management valued the individual mechanics and treated them as autonomous professionals. The fact that all AMTs hired held A&P certificates reinforces this observation.

Aloha's management appeared to hold to the traditional view that AMTs are fully skilled and capable and require little or no direction or professional support. This attitude is consistent with several of the AMT characteristics listed previously (Table 2.2): self-sufficiency, integrity, desire to think things out on their own, and so forth. It also implies that AMTs and management alike saw little need for communication with each other.

Unfortunately, many of the company's AMTs in 1988 were young and relatively inexperienced with the B-737. Although the longer-service AMTs had lots of experience, they had not been provided much recurrent training by the company and were considered below industry standards (observed in the Boeing report included in the NTSB findings [Ref. 1 p. 234]).

Some AMT characteristics from Table 2.2 (not willing to ask for help, not sharing thoughts too frequently) could prevent management from appreciating this situation. Likewise, the industrywide practice of mostly top-down communication does not invite valuable communication in the other direction via observations, suggestions, questions, and answers. Crucial communication between AMTs and management is much less likely in such a low-communication culture.

The Company's Larger System

Now let's look at various aspects of the company's larger system (from Table 3.1). These issues must also be included in order to better understand the situation at the time of the accident. Naturally, it's a mixed bag.

TABLE 3.1 ALOHA AIRLINES MAINTENANCE OPERATIONS, 1988

Positive Aspects of Maintenance	Negative Aspects of Maintenance
◆ Maintenance hires all AMTs with A&P certificates.	◆ Two-thirds of maintenance force is young and unfamiliar with B-737.
◆ One-third of maintenance force is very experienced.	◆ Few had recurrent training and most are technically below industry standards.
◆ Maintenance mgt. places trust and reliance on AMTs as individual professionals.	◆ Maintenance mgt. provides little technical support to AMTs—FAA Airworthiness Directives go straight to hangar without interpretation.
◆ AMTs receive good pay and benefits for long service—valuing the individual wage-earners.	◆ Senior hourly employees enjoy unusual influence in middle mgt. decisions and are resistant to change.
◆ Labor relations harmony. No strike history.	◆ Evaluators noted lack of motivation and *esprit de corps*.
◆ Company has strong community relations and local goodwill.	
◆ On-time reliability very high (99.6%).	◆ The best, most experienced AMTs are assigned to flight line and contract maintenance.
◆ Contract maintenance customers very satisfied with support received.	◆ Overhaul AMTs are reassigned to contract maintenance on almost daily basis. Sometimes that overhaul group is reduced by two-thirds.
◆ Cost-effective: 14 maintenance personnel per aircraft (compared to 27 nationally).	◆ The regular complement of AMTs assigned to overhaul is insufficient to complete checks in a reasonable amount of time.
◆ Cost-effective: Maintenance is 6.6% of operating expenses (compared to 11.2% nationally).	◆ Overhaul cases are known where areas have been painted over or closed up before final inspection.
◆ Cost-effective: Little aircraft flight time is lost to "D" Checks because it is performed in 52 small segments over several years.	◆ With many small segments in the "D" Check, an aircraft's interior is usually not removed. It is difficult to keep a clear idea of overall aircraft condition.
◆ Fleet is all B-737-200s. This homogeneity simplified maintenance operations and support.	◆ Fleet is old (average age 13.1 yr compared with 12.7 yr for all types U.S. fleet nationally) and has the second highest cycle 737s in the nation—they are in deteriorated condition.
◆ Routes flown at lower altitudes and pressure differentials than most U.S. airlines puts less strain on aircraft.	◆ Short flights and many pressurization cycles place added strain on aircraft.

Sources: NTSB [Ref. 1]; Donoghue [Ref. 3]; Ramirez [Ref. 4]

The company did fit into its local community. Its relationship with its employees, their community, and their trade union were all positive. There was no record of employees refusing work on safety grounds.

Customer service was obviously important to the company's management and employees. Satisfied passengers are especially important on a short-haul airline like Aloha. The "other" customers—those airlines that Aloha serviced with its maintenance staff—also indicated they were very satisfied with Aloha's work. The company paid attention to what its customers wanted and placed its AMTs in position to deliver.

AMTs put in the effort and the hours, but, at least in the overhaul area, they lacked the resources, the skills, and even the encouragement to do excellent work.

The importance of low-cost operation is abundantly clear from Table 3.1. Fewer AMTs per aircraft, lower maintenance costs as a percent of operating expenses, and management of heavy maintenance (or overhaul) for a minimum loss of aircraft flight time are all important ways to minimize labor costs and maximize return on capital investment. The right-hand column of Table 3.1 suggests, however, that this twin emphasis on customer satisfaction and heavy cost constraints directly affected the time and personnel available for heavy maintenance and overhaul. The younger and less-experienced AMTs assigned to overhaul were struggling to maintain aircraft properly.

The fact that the company's fleet was relatively old with high cycles (take-offs and landings) is also a sign of tight money in management's investment strategy. The overhaul group had lower priority and could fully inspect the aircraft only rarely.

It was also the company's bad luck that 70% of its B-737 fleet was of the early "cold-bonded" construction, which was, according to Airworthiness Directives (ADs), more likely to disbond and was thus weaker at the lines of rivets on the belt line lap seams that would ultimately fatigue and break [Ref. 4; 5].

The deck was stacked against the overhaul group and against safety.

Despite the significant role that the overhaul group played in the company's low-cost operation, their accomplishments were unrecognized and unsung.

For instance, management routinely sent FAA ADs and service bulletins straight to the hangar floor where AMTs had to figure out by themselves what to do and how to do it. And like many smaller airlines in 1988, the company had no Engineering Department to help AMTs comprehend and interpret complex ADs or to provide oversight of inspector training and performance.

The fatigue-cracking scenario was not carefully addressed. Crucial company inspection records were not completed, and some terminating actions specified in the ADs were done incorrectly [Ref. 1 pp. 26, 44, 53, 56, 62].

Communication and Collaboration

Customer priorities were clear, and the best of the company's AMT workforce were assigned to departments that served those priorities. As an airline that operated mainly during the day, the usual "social distance" between flight line and hangar was further lengthened by assigning high-status AMTs to the day shift and lower-status AMTs to the night shift.

In any company, in any industry, the social distance between high and low priority work, time distances between day and night work, and physical distance (e.g., between flight line and hangar) all combine to make communication between departments and shifts difficult. Such "human factors" are all too often given little thought.

In addition, management's failure to provide a system for technical and communication training for mechanics and inspectors resulted in lower-quality work in overhaul. For its own part, overhaul resisted attempts to change or correct this situation [Ref. 3]. This work requires high levels of communication and coordination, but the structure, skills, leadership, and motivation for open communication were not up to par. Even if everyone knew that open communication was key, they lacked the skills, the permission, or the comfort to confront the situation and solve it.

Whether such communication barriers existed elsewhere in the industry was a matter of grave concern. The NTSB had said quite plainly this was an industrywide problem. Considerable interest in the topic was demonstrated by the attendance numbers at public conferences on "aging aircraft" and "human factors in maintenance" during 1989–90. Still, many airline maintenance managers were hopeful that the misfortunes of one small airline in a remote location didn't really reflect their management or their communication practices.

Then disaster struck again.

The Eagle Lake Accident

In late 1991, an airplane crashed after losing a section of its tail in flight. The crash was directly attributable to maintenance operations.

All 14 people on board perished (11 passengers and a crew of three).

At about 10 a.m. on September 11, 1991, an Embraer 120 aircraft, on a regularly scheduled flight from Laredo to Houston, Texas, lost radio and radar contact with the Houston air traffic control center (ARTCC) as it began its descent for approach.

Eyewitnesses on the ground near a cornfield at Eagle Lake, where the plane crashed, described it as suddenly consumed by a fireball and pitching over. A bright flash followed, then the left wing separated, and finally the plane spiraled downward in a hail of falling parts.

The NTSB cited failure of the carrier's management (Continental Express Airline) to ensure compliance with approved maintenance procedures as a contributing cause of the crash. Specifically, a scheduled maintenance operation started on one shift had been carried over to the next shift without being adequately communicated [Ref. 6].

Outline of the Accident

It was determined that the Eagle Lake crash resulted from an in-flight separation of the leading edge of the left horizontal stabilizer from the airframe. This reduced the down force of the stabilizer (the horizontal tail plane) and caused the aircraft to pitch down violently for a few seconds, then roll to the right and disintegrate.

Both the left and right horizontal stabilizers had been scheduled for deicer boot replacement at the Houston hangar the night before the accident, and work had begun on both. Only the right side was completed, however.

On that night, according to the Safety Board's report, one-fifth of the company's second- (evening) and third-shift (midnight) hangar workforce had been involved in work on the tail section of that aircraft. Both second-shift supervisors on duty that night were involved in the repair of the aircraft in question, and they were clear about which of them had responsibility for it throughout the shift.

The third-shift hangar supervisor arrived more than 45 minutes before his shift began, and, in the course of familiarizing himself with the night's schedule, asked one of the second-shift supervisors if work had started on the aircraft's left stabilizer. He was told "no."

The third-shift supervisor then told the second-shift supervisor that he would be able to change the right side boot, and that the left side boot could be changed on another night. This dialogue was held with the second shift supervisor who was *not* responsible for that aircraft.

The safety board determined that a second-shift AMT inspector had removed the screws securing the top of the deicing boots on both the right and left stabilizers of the "T-type" tail. His AMT (mechanic) workmates had removed the screws on the lower side of the deicing boots only for the right side. The top surface of the stabilizers on this aircraft are some 20 feet above the ground; thus the missing screws on the upper surface were not readily visible.

Here are the makings of a real communication nightmare. It's worth your while to read the next few paragraphs carefully.

The second-shift AMT (inspector) who removed the screws on the upper surface had been disciplined for inspections on two occasions prior to the accident. One incident involved not finishing all required paperwork.

An incoming third-shift AMT mechanic arrived early before relieving the second-shift AMT who had removed the screws. The incoming AMT was given a verbal turnover by a mechanic working on the job, who later reported that he explained that all screws were removed from the right side and only the screws from the top surface had been removed from the left stabilizer. The incoming AMT also reviewed the written turnover sheet and then went to the break room to wait for the start of his shift.

Unfortunately, the outgoing second-shift AMTs had not yet completed their written reports. The second-shift AMT inspector, who had removed the upper surface screws, left those screws in a bag on the manlift and then wrote in his Inspection Department's written report, "helped the mechanic remove the deice boots." He then clocked out and left for home.

The list of communication problems is not yet complete.

Although verbal turnover was accomplished on the removal of the right side deicer boot, it is unclear whether the second-shift AMT failed to report the missing screws on the left stabilizer, or whether the third-shift AMT failed to hear it. To make matters worse, the third-shift AMT who received the verbal turnover was actually assigned to another aircraft. That third-shift AMT relayed his message to a fellow third-shift AMT who arrived at the hangar floor.

Yet *another* third-shift AMT arrived at the hangar and was informed by the third-shift hangar supervisor that he should talk to the second-shift supervisor to find out what had been accomplished. There was no discussion regarding *which* of the two second-shift supervisors the third-shift AMT should talk to. He talked with the supervisor who was *not* in charge of the aircraft in question. The AMT asked if any work had been performed on the left deicer boot. Excellent question. The answer? The second-shift supervisor informed him that he did not think he would have time to change the left deicer boot that night. That's it.

Third-shift AMTs removed the right stabilizer deicing boot in the hangar. The aircraft was then pushed outside to make room for another aircraft. Work on the horizontal stabilizer resumed in the dark without flood lights. The right side leading edge assembly (including the deicing boot) was reinstalled outside.

In the morning the airplane was cleared for flight. The first flight was from Houston to Laredo at 7:00 a.m. There is no evidence from the morning's preflight that the flight crew knew about any of the work performed on the horizontal stabilizer, and there was no policy or regulation that would require them to know. Although that first flight was officially without incident, a passenger later reported that during the flight he had asked the flight attendant to give him a new seat because he had been awakened by vibrations. The passenger did not inform the flight attendant or flight crew of the vibrations, however. Nor did other passengers on that flight report vibrations. The vibrations could have been the result of the increasing looseness of the left horizontal stabilizer's leading edge assembly (including its deicing boot).

The airplane crashed on its next flight.

Collaboration, Communication, and the Eagle Lake Accident

Despite the horrible outcome of their work on that plane, the maintenance people at Continental Express had shown positive motivation and collaboration in their work. Although commendable and needed, these are not enough.

Let's analyze this case a bit further.

AMTs and their supervisors had swarmed the aircraft over two shifts to make repairs to its tail section. But communication didn't keep pace with the work. They wanted to help one another, but didn't know how to work together effectively.

The work cards for the deice boots replacement job were not issued to second-shift AMTs by their supervisors. That package of cards was originally assigned to the third shift for completion, but the second-shift supervisors elected to start work on the deice boots to assist the third shift with the workload. Because the work cards had not been issued to the second-shift AMTs, no entries were made on them that would alert the third-shift supervisor and AMTs that work had been started on *both* the right and left deice boots.

The two second-shift supervisors shared the work direction, but one of them received requests for information from third-shift employees that he was unable to answer knowledgeably or accurately.

The third-shift supervisors and AMTs contributed to the confused communication by being on the job well before the second-shift personnel had a full picture of their own accomplishments clearly described in the turnover record and aircraft log book.

Third-shift AMTs did not take the time to identify who had the accurate information about the status of the work. Likewise, eager members of the third shift arrived for work so early that the outgoing AMTs didn't yet have a clear idea of what they were turning over when information was sought. Verbal turnovers took place, but those receiving them were not always the ones assigned to the work.

All of this is in addition to the second-shift AMT who misinterpreted the confused work direction he received from the two second shift supervisors. He left the screws behind, he left a vague written report only intended for other inspectors, and then he left for home.

The final opportunity to inspect the work was stymied by ambiguous wording and interpretation of the maintenance manual regarding whether deicing boots were a "required inspection item" (RII) since they were attached to the horizontal stabilizers, which are RII. The company's managers later confirmed that, *in practice*, they were not RII.

Mistakes in shift turnover are among the worst fears in maintenance. Usually there are plenty of opportunities to review, check, confirm, and verify that the work has been accomplished correctly. These quality control activities, however, come after the fact—they don't create error-free work in the first place, and they don't often find errors that they weren't looking for. The death-dealing deicer fiasco is a case in point.

"Fail-safe" is built into the professional attitude of all maintenance people and it's built into the pace of the work, the paper trail, the policies, and the regulations. It is a worthy ideal that maintenance pursues with great energy and intelligence.

There is no truly fail-safe system, technical or human—never was, never will be. It is possible, however, to get closer and closer to that vision—less and less chance of failure—even as our technical and human systems get more and more complicated. Later chapters will illustrate how other industries *are* getting much closer to perfection by designing their high-tech systems in tandem with a high-communication approach. And the Eagle Lake accident might have thus been avoided.

Shift turnover rarely leads to tragedy. But rarely is not good enough. And better communication in maintenance would certainly have prevented 14 deaths.

The Eagle Lake accident could have been ignored by the aviation maintenance community. Fortunately, it set off an industrywide alarm: What happened at Eagle Lake could happen to us.

The accident increased industry awareness that better communication among AMTs, between shifts, with management, with other divisions in maintenance, and with other departments of the parent company is critical.

The industry learned, again, that communication errors can and do cost human lives. And they learned to look at the communication system itself for answers. As the NTSB accident report concludes: "The Safety Board does not believe that the maintenance issues [in this case] were related solely to the actions of individual employees who were in the hangar the night before the accident. The failure to follow proper turnover procedures suggests a general disregard for following established procedures" [Ref. 6 p. 43].

The accident investigation revealed that maintenance people were generally aware of the correct procedures. One NTSB Board Member went so far as to say that the company's *"corporate culture* led to the maintenance breakdown that caused the crash" [Ref. 7, emphasis added].

Everyone in the system caused this crash. But the story continues.

Culture Is Hard to Change

In December 1992, 14 months after the Eagle Lake crash, and almost 6 months after the NTSB report was released, a remarkably similar event occurred at the Continental Express Houston maintenance base. Shift turnover problems resulted in an aircraft taking off with missing screws that were removed during overnight maintenance. The pilot returned to the Houston airport after he felt vibration and saw a panel flapping on the left wing. The company fired a mechanic and an inspector involved in the case.

NTSB investigators found that 12 out of 51 screws were missing from the panel and two more screws had broken off because of the vibration [Ref. 8]. According to an NTSB board member, many of the shift turnover processes used at this airline are in use by most airlines in North America [Ref. 9]. Furthermore, this company's reported way of handling the problem through punishing personnel illustrates the old management approach.

A Parting Thought

Four questions arise from the cases in this chapter:

1. Why doesn't the communication process always work when there are plenty of checks and balances?
2. Isn't the problem that the processes aren't adhered to?
3. Is this also true at other airlines?
4. Why is the learning so slow?

Here are the short answers—more to come in later chapters:

1. The process design does *not* have all the right checks and balances that may be needed. The system needs to be redesigned.
2. If the process isn't being adhered to, first consider the process design itself to be faulty, not the individual. Understanding and compliance issues must be included in the design. Employees are not to blame when the system makes it hard for them to understand and comply with expectations. The process design needs to be improved.
3. Yes. Other airlines have similar problems with communication.
4. It's not difficult to understand why we're so slow to learn from others' mistakes when we don't learn well from our own.

References

[1] National Transportation Safety Board (NTSB). "Aircraft accident report: Aloha Airlines, Flight 243, Boeing 737-200, N73711, near Maui, Hawaii, April 28, 1988." NTSB/AAR-89/03, Washington, D.C. (1989).

[2] National Transportation Safety Board (NTSB). "Aircraft accident report: Eastern Air Lines, Inc., Lockheed L-1011, N33EA, Miami International Airport, May 5, 1983." NTSB/AAR-84/04, Washington, D.C. (1984): 50.

[3] Donoghue, J.A. "After the storm, Aloha in calm waters." *Air Transport World* (July 1990): 70–75.

[4] Ramirez, A. "How safe are you in the air?" *Fortune* (May 22, 1989): 75–79.

[5] Federal Aviation Administration (FAA) Airworthiness Directive 88-22-11 (and associated Boeing 737 Service Bulletin 737—53A1039), Washington, D.C. (October 27, 1988).

[6] National Transportation Safety Board (NTSB), "Aircraft accident report: Britt Airways, Inc., dba Continental Express Flight 2574, in-flight structural breakup EMB-120RT, N33701, Eagle Lake, TX, September 11, 1991." NTSB/AAR-92/04, Washington, D.C. (1992).

[7] Associated Press, "Missing screws caused plane to crash, safety board says." *Orange County Register* (July 22, 1992).

[8] Sharn, Lori. "Missing screws: Chilling similarity to fatal flight." *USA Today* (December 23, 1992).

[9] Goglia, J. "Panel presentation on airline maintenance human factors." In *Proceedings of the 10th FAA Conference on Human Factors in Maintenance* (1996).

Chapter 4

A Short History of Communication and Human Factors Management in Aviation Maintenance

OVERVIEW: Human Factors Management Is a Powerful New Science in Aviation

Charles Edward Taylor (1868–1956) was the mechanic who worked with the Wright brothers on their initial flights at Kitty Hawk in 1903. He was the first American aviation mechanic in powered flight. He not only kept that first engine going, he designed and built it. Taylor continued to develop and build engines for the Wright brothers' early planes, and he maintained them.

In 1912, Taylor joined the newly formed Wright-Martin Company to manage the engine shop. He remained there until 1920. In 1965, Charles Edward Taylor was posthumously inducted into the Aviation Hall of Fame in Dayton, Ohio.

At the beginning of manned flight, engineers and mechanics made airplanes airworthy and pilots made them fly. They were all crucial to the success of the fledgling industry, and their successes were magnificent. But their attention to the possibilities for human error and the possible consequences held a relatively low priority.

That has long since changed. And today human factors science is the cutting-edge strategy being used to reduce human error in maintenance and elsewhere. Many airlines are implementing parts of these programs, often called MRM (for Maintenance Resource Management). MRM aims to improve work performance by improving workplace communication.

What Human Factors Management Is All About

It's about workplace communication systems.

According to definitions found in recent FAA literature, the term *human factors* denotes a multidisciplinary field devoted to optimizing human performance and reducing human error. It incorporates the methods and principles of the behavioral and social sciences, engineering, and physiology [Ref. 1]. For our purposes here, "human factors science" is the applied science that studies people working together and in concert with tools and machines. Human factors science embraces variables that influence individual performance and variables that influence team or crew performance. Thus human factors management is the art and science of building and maintaining effective workplace communication systems.

However, even the latest official definitions of human factors management do not adequately encompass all the improvements required in aviation today. Communication and collaboration must embrace an even wider context. Beyond the individual and crew, human factors management must be understood to include the entire organization—that complex we call the "socio-technical system" and its culture.

In their recent book, *Beyond Aviation Human Factors,* Daniel Maurino and his colleagues challenge traditional human factors management to go beyond their "focus on the individual human operator—mainly pilots, controllers or mechanics . . . as the sole professions responsible for safety in aviation" [Ref. 2 pp. ii–iv]. These authors propose an organizational viewpoint that takes a new approach to safety, going beyond the individual professional, the specific regulation, and the newest technology to encompass all elements as a whole system. Such a system includes maintenance management, the larger company, and even the wider international air travel system. Picture concentric circles, with each ring affecting all the others.

There is plenty of potential in this way of thinking. It promises to do more for reducing human error than any individual or one-on-one approach could. It recognizes that errors may come from any part of the system, at any time. To design a workplace communication system that keeps errors in check regardless of their source offers real benefits for the present and for the future—*fewer* errors just when growing complexities would ordinarily have generated *more* errors.

To capture these benefits will require significant change in the way maintenance work is organized and managed. And to do this will require changing pretty much everything.

For an industry that hates to gamble, this "bet the pot" proposition must seem risky. But the risk is both minimal and manageable. Properly approached, such an organizational shift would produce a work system that is safer, more responsible, and more economical than the old-style organizations in place today. Far-sighted organizations around the world have proven this repeatedly—this is no mere hypothesis [Ref. 3].

Aviation maintenance already understands this concept on an intellectual level. But an operational shift in focus has not yet occurred. There's been a lot of talk about using the insights of human factors management, communication, and all the rest. As for actually doing it, though, much of aviation maintenance is keeping one foot on the sidelines, wondering whether it could or should actually get into the game. And some of those who've already introduced MRM or some other stab at communication improvement still have doubts and still withhold their full commitment to change.

Despite their reservations, however, a refocused, better organized work system is well within the grasp of any airline whose management is willing to learn, to lead the way, and to stay the course. Who could argue, really, with the goal of reorganizing workplace communication to improve safety?

The Evolution of Human Factors Management

1914–1918: Human Factors Defined As Individual Abilities and Skills

The human factors involved in flight have been examined from the beginning. World War I, as the first mechanized war, used the airplane in substantial numbers and in a variety of ways. It was important to attract airmen who would fly the new contraptions. These people were first of all expected to be fearless. It didn't much matter what else they were. The primary method of selection for this personality characteristic was self-selection. The first pilots were volunteers. Intelligence was the next important human factor considered. Intelligence tests developed during World War I became the most used selection tool of the time.

Once a sufficient number of applicants with the noble virtues of fearlessness and intelligence were found, technical proficiency then took precedence. The courageous young cadets were trained to operate their assigned machinery with knowledge and skill.

During flight training it was noted that even with the best selection available at the time (or now for that matter), a near miss or a nonfatal accident could cripple a cadet's ability to fly. Courage was reinstated by simply putting the cadet in the air again as soon as possible. Training pilots to *control* their fear surfaced as the next important human factor, defined as self-awareness and emotional control. After the war and for the next 20 years, these core factors guided pilot selection and training.

Another factor, human fatigue, received serious official attention in British munitions plants during World War I. In their efforts to improve wartime production, the British learned lessons about using rest and break times, but these lessons were not incorporated into aviation practice until many years later.

1939–1945: Human Factors Defined As Machine Design and Crew Coordination

As World War II opened in Europe, the allies urgently needed to achieve overwhelming air superiority. To do this, developing efficiencies in flight operation became crucial. Significant breakthroughs were made in adjusting the machine to the man (human engineering or ergonomics) as the Army Air Corps redesigned cockpit instruments and other equipment for more effective use.

England's flight simulator (the "Cambridge cockpit") was another significant development. Pilots could now be trained, their skills and reactions tested without endangering either human life or an aircraft. Intelligence tests and psychomotor tests were widely used in pilot selection by this time, and they were also used to select people for other military aviation occupations, including mechanics.

The importance of close teamwork among bomber crew members was also recognized. Flight crew management was studied quite extensively during World War II. Much was learned and applied as a result of those studies about group dynamics in these crews and the management of professional communication during high-stress periods.

There were no similar studies on aviation maintenance operations during the war years or for a long time after. Thought of as individual contributors, mechanics were rarely considered part of the larger crew. If anything, mechanics were trained and cautioned to be wary of others outside their vocation. A passage from the 1941 "Mechanic's Creed," from the U.S. Civil Aeronautics Board (CAB), reads as follows: "*I* pledge *myself* never to undertake work or approve work which *I* feel to be beyond the limits of *my* knowledge, nor shall *I* allow any non-certified superior to persuade *me* to approve aircraft or equipment as airworthy against my better judgment . . . " (The added emphases show the highly individualized focus of responsibility at the time. This has not changed much in the five decades since then.)

1950–1978: Human Factors from Korea to Vietnam and Beyond

For the next 30 years, human factors management in aviation continued to progress. Studies of pilot selection, simulator training, and human engineering in cockpit design continued after World War II. But most of what had been learned about group dynamics and communication from the bomber crew studies was subsequently forgotten in military aviation, air travel, and commercial air freight operations.

The model of the individual airman, rather than the collaborative system model, remained in ascendancy in both civilian and military aviation. And studies of individual human factors were undertaken in military aviation maintenance during the 1950s. Interest in man-machine interactions, for instance, led to examination of human factors in equipment design, which aided subsequent selection and training of aviation mechanics [Ref. 4; 5].

During the Vietnam War, the quest for greater aviation safety brought about a systematic approach to error reduction. A significant contribution was made by human factors studies in military aviation and in aircraft manufacture during the Vietnam War. Zero-defects quality programs, for example, were introduced during the 1960s with varying degrees of success in reducing errors. As a part of the zero-defects approach, maintenance errors were also addressed in the 1960s studies. One elaborate error-cause analysis at the time listed over 50 categories of maintenance errors, at least 10 of which involved human communication of some kind [Ref. 6].

The 1960s was also witness to the "crackdown" approach to human motivation and behavior control. "Crackdown" programs were basically one-way communications from management, with behavior control based on fear and punishment. These tactics were often used simultaneously with zero-defects programs for error reduction. Did they work? As one author wrote at the time: "Such crackdown drives errors into hiding—which often only delays the problem in which errors manifest themselves during a later and more critical stage of an operation" [Ref. 7].

To avoid seemingly inexplicable, unpredictable human mistakes, a systemic approach to human factors management introduced during the 1960s and 1970s embraced the ideal of "foolproof" equipment designs in addition to the continuing exhortations for zero defects. Other management efforts to reduce errors addressed motivation. Some military management studies during the 1960s and 1970s examined the effects of positive, rather than negative, motivators. These results suggested that, while motivation due to good morale often improves maintenance safety performance, motivation produced from negative sources (such as fear, guilt, or ridicule) seldom achieves the same effect [Ref. 7]. "Modern" management texts in the 1960s thus challenged the crackdown method [Ref. 8]. A debate raged over whether organizational effectiveness is better served by autocratic or democratic means. By the late 1960s, participative management ideas were making great inroads into a number of U.S. industries; however, few airlines, and fewer of their maintenance departments, looked to these more participatory methods.

A small backlash against the punishment-centered approach to improving worker motivation and performance in aviation occurred, and the 1970s saw some use of "positive reinforcement" to achieve desired behavior. Such programs were founded on the psychological theories of B.F. Skinner and focused on the use of rewards (and the disuse of punishment) in modifying individuals' behavior [Ref. 9]. In commercial aviation, the best-known application of positive reinforcement was Emery Air Freight's program of Skinnerian Behavioral Modification for ground crews and package loaders, widely cited in the management literature following the 1973 study of three years of productivity improvement in that company [Ref. 10 pp. 286–7]. Both the publicity and the program have since passed from the scene.

The Airline Deregulation Act of 1978 and the Decade to 1988

Before 1978, the airline industry was regulated through the Civil Aeronautics Act of 1938. Among other things, the 1938 legislation produced a peaceful market for air travel and rationalized routes and fares. But it also allowed wasteful management practices and wages that were higher than other industries.

The Deregulation Act of 1978 brought with it "the best of times and the worst of times." For airlines prepared for the opportunities as well as the chaos, it was a time for doing business competitively, with routes and fare structures designed for profitability.

A new breed of airline executive came to the industry and brought with them the skills to deal with the hard-nosed market environment. The new airline CEO was a businessman, often with an MBA, often with little prior experience with airplanes or flying. These managers would determine not only what to charge or where to fly, but would experiment with different kinds of airline service and style.

As competition strengthened some airlines, it likewise weakened others. The weak ones were ready takeover targets. In some cases, the executives newly in charge of these airlines were themselves takeover kings and merger giants—often self-made men who were skilled primarily in finance, greenmail, and leveraged buyouts. They were less skilled in managing a technical, engineering-based enterprise. But irrespective of the interests and skills of its senior executives, the industry became less "technical" and more of a "commodity."

Where management did not change and where established carriers did not modify their well-entrenched systems, small carriers rushed in. These start-up companies quickly created new systems from scratch and they fully explored available business possibilities. Two innovative airlines that came into being after deregulation were Southwest Airlines and People Express.

Deregulation also brought a flurry of airline mergers and acquisitions that has continued to the present. These mergers often permitted economies of scale and more efficient use of fixed resources, and they allow short-term revenue improvements through the sale of excess capital equipment. With mergers, redundant staff support units—including accounting, human resources, marketing, and engineering—can be greatly reduced yet continue to provide adequate services.

During the early 1980s, airline Maintenance Departments also experienced reduction pressures from mergers, and these reductions were "justified" by the excellent fleet reliability statistics of that period. With airplanes needing so little repair or attention, fewer mechanics were required and immediate savings in labor costs could be shown.

Prior to deregulation, airline engineering departments worked with aircraft manufacturers to design airplanes for their special needs. After deregulation, these departments were radically reduced in size and their involvement in designing new aircraft diminished in proportion. New ways of doing airline business emerged, such as acquiring aircraft off the shelf and using leased instead of purchased aircraft. Keeping aircraft longer rather than acquiring new equipment became a competitive strategy in the early 1980s. If equipment could be leased instead of owned, so too could employees be "leased." Outsourcing hangar maintenance and heavy overhaul became increasingly popular among airlines. Smaller airlines and airlines with expanding routes found contracting even line maintenance to outside companies to be an efficient business strategy.

Prior to deregulation, and through 1987, human factors management in U.S. aviation maintenance remained in its 1960s mode. A comprehensive review of the literature published between 1976 and 1987 shows little about aviation maintenance workers. Of 50 articles published during this period, only 15 even mention maintenance, and these deal with the physiology of human response. A few studies discussed the whole person in context [Ref. 11; 12]. One reference describes and reports technical advances in military aircraft engine design that supposedly made field maintenance duties "soldier-proof" [Ref. 13]. Eliminating the human factor through technology (or at least as much as possible) is in direct collision with the notion of a system of informed decision-making and cooperation. One paper [Ref. 14] suggests more management controls (crackdowns) in naval aviation maintenance.

Culture Clash in Aviation Maintenance

Management texts in the United States of the 1960s challenged the crackdown method [Ref. 8]. In the subsequent decade, alternatives to using negative motivators in aviation were explored by the U.S. military and by one airfreight company [Ref. 7; 9]. In the 1970s and 1980s, maintenance departments in other countries' airlines began experimenting with participatory management, and they reported success.

Human Factors Management in Other Countries

Two of those 15 maintenance articles published between 1976 and 1987 [Ref. 15; 16] reported positive results at Air Canada and at British Caledonia Airlines using discussion groups of mechanics to solve technical problems. In addition, Scandinavian Airlines and Japan Air Lines widely publicized their high-involvement management approaches during 1987 and 1988, and they described positive effects on maintenance operations. But whether the airline was British, Canadian, Scandinavian, or Asian, such experiments with greater involvement by (and communication with) maintenance employees were all searching for excellence outside of America's deregulated skies. Too bad.

In America at the time, such ideas were considered too foreign and too costly. Managers at most U.S.-based airlines paid them little attention.

The Short History of People Express

One rather famous exception to this widespread neglect of modern management techniques was the People Express airline. Formed in 1980 specifically in response to the 1978 deregulation act, People Express was non-union and 33% employee-owned. The employees rotated jobs and filled in where they could when time permitted. They were involved in every phase of the company's operations. People Express was also a low-cost airline, spending about five cents per available seat mile [Ref. 17]. They were successful, and they exemplified a new management culture in U.S. commercial aviation. But this new culture was soon to clash with the old: In 1987, People Express was acquired by Continental Airlines.

Texas Air Corporation and Continental Airlines

Continental had responded to deregulation quite differently than People Express had. In 1981, Texas Air Corporation acquired Continental Airlines. The culture of the new company was traditional. Three years later, in October 1983, Continental Airlines' declaration of bankruptcy was widely suspected to be a strategy to negate labor contracts, thus using Chapter 11 to become a lower-cost airline. The company's declaration of bankruptcy coincided with the abrogation of Continental's union contracts, which was followed by radical reductions in both wages and manpower [Ref. 18]. Given this behavior, the company culture of Texas Air/Continental moved from merely traditional to excessively harsh and severe. Thus, before their 1987 merger, People Express and Texas Air/Continental Airlines marked the extremes of human resource practices in aviation management.

During the 1980s, other airlines also reduced their maintenance workforces, through furlough or attrition, and wage concessions were frequently requested of maintenance staff. For many in maintenance it seemed the worst of times indeed.

Deregulation and Human Factors Management at Eastern Airlines

In 1984, Eastern Airlines, another well-known name among U.S. airlines, applied social science to its labor relations. Two threads of collaboration came together in this program [Ref. 18]. First, labor and management agreed to wage concessions and control concessions, and second, a program of employee involvement and participative management was begun. Too little, too late? With former astronaut Frank Borman in charge and despite high hopes, Eastern was acquired in 1986 by Texas Air Corporation (who also owned Continental, remember?). The buyout was the end of Eastern Airlines as employees had known it [Ref. 18].

By 1987, Eastern and People Express had both lost their promising high-involvement cultures through acquisition by Continental with its strongly autocratic management culture.

Cultural Change in Maintenance in the 1990s

Evidence from aviation maintenance collected during the 1990s reveals practical applications of systems thinking and cultural change taking the form of changes in management behavior, in jobs and organization structure, in strategy and policy, and/or in values. And virtually all of these changes involved communication and collaboration among people.

Following the succession of its CEO, the culture at Continental Airlines began to focus on improving and opening communication. In 1991, Continental received wide publicity when they embarked on CRM-type training in maintenance [Ref. 19]. "CRM" stands for Cockpit Resource Management, the forerunner to Maintenance Resource Management (MRM). At the 1992 FAA meeting on Human Factors in Aircraft Maintenance and Inspection, John Stelly described the first year's positive effects of communication training for his airline's more senior technical operations personnel. Improvements in ground damage and occupational injury rates were reported as positive outcomes of the training [Ref. 20]. At the 1993 FAA meeting on this topic, three speakers described the positive effects of new task allocation and communication among maintenance crew members in their respective airlines.

A change in organization and job design of inspectors at United Airlines was reported, in which inspectors remained accessible during the entire period of heavy maintenance and overhaul, stayed in closer communication with mechanics during repairs, and "bought back" their own initial nonroutine defects [Ref. 21]. These changes resulted in meeting overhaul schedules, fewer air turnbacks, and higher subsequent quality.

A strong and clear organizational culture created and sustained by Southwest Airlines' CEO, Herb Kelleher, resulted in improved communication between maintenance and other departments [Ref. 22].

At TWA, a new program for communication between the mechanics' trade union (I.A.M.) and maintenance management improved quality control [Ref. 23].

In another cooperative effort with the IAM, USAirways and the local FAA office (FSDO 19) achieved notable successes in reducing errors in maintenance documentation by developing new channels of open communication among AMTs, their maintenance foremen, and maintenance management [Ref. 24; 25]. Chapters 8 and 9 will show that this interest in improving performance in maintenance paperwork is not trivial.

Beyond Human Factors Management

In a 1996 speech to the Ninth FAA Conference on Human Factors in Maintenance, John Goglia, an NTSB member and former airline mechanic, argued for moving the focus of human factors management from the individual to the larger maintenance system. In distinguishing Maintenance Resource Management (MRM) from other ways of applying the human factors involved in maintenance, Goglia stated that terms like "technician resource management" placed undue attention on the individual AMT or work crew [Ref. 26]. Goglia said that "the individual may be blamed or found to be at fault, when experience has shown that the fault is [frequently] systemic."

This point of view is shared by the authors of this book, and the case studies presented here support this view. Maintain high standards for individual performance, but broaden your outlook to better manage the impact of the larger system. Many seemingly individual errors are actually produced by a poorly structured system of communication and social support.

The golden road to improved maintenance safety and performance focuses on the dynamics of the whole system—including aircraft characteristics, tools, physical work environment, technical skills and knowledge, and human physiology—as always. But a shift in focus is necessary to better address the issues of communication and collaboration within and between groups, among departments, and with separate organizations (such as regulators, airlines, repair stations, trade unions, and special interest groups). This broader communication net brings more attention to bear on the whole culture that guides the behaviors of related departmental, corporate, national, and international bodies.

The new mindset is more like that of leading an effective government than managing a corner repair shop. Even those closest to that kind of ideal thinking still have a fair way to go.

To optimize both the technical system and the social system for better results requires a careful and participative analysis and redesign of the way each organization is structured and managed. The concepts and techniques for doing this are a well-developed part of human factors management called "socio-technical system design," emphasizing communication, teamwork, and accountability.

References

[1] FAA, "Crew resource management training." Advisory Circular 120-51A. Washington, D.C. (1993).

[2] Maurino, D.E., J. Reason, N. Johnston, and R.B. Lee. *Beyond Aviation Human Factors*. Aldershot, Hampshire: Ashgate, 1995.

[3] Naisbitt, John, and Patricia Aburdene. *Re-Inventing the Corporation*. New York: Warner, 1985.

[4] Miller, R.B. "Anticipating tomorrow's maintenance job." *Human Resources Center Research Review 53-1*. San Antonio, TX: Lackland AFB (1953).

[5] Gagne, R.M. "Methods of forecasting maintenance job requirements." In *Symposium on Electronic Maintenance*. Washington, D.C. DoD R&D (PPT 202/4) (1955): 47–55.

[6] Willis, H.R. *Human Error and Safety Program*. Denver: Martin Marietta Co., 1967.

[7] Cornell, C.E. "Minimizing human errors." *Space/Aeronautics* (March 1968): 72–81.

[8] Likert, R. *New Patterns of Management*. New York: McGraw-Hill, 1961.

[9] Skinner, B.F. *Contingencies of Reinforcement*. New York: Appleton-Century-Crofts, 1969.

[10] Saal, F.E., and P.A. Knight, *Industrial/Organizational Psychology*. Pacific Grove, CA: Brooks/Cole, 1988.

[11] Lock, M.W.B., and J.E. Strutt. *Reliability of In-Service Inspection of Transport Aircraft Structures*. CAA Paper 85013. London: Civil Aviation Authority, 1981.

[12] Strauch, B., and C.E. Sandler. "Human factors considerations in aviation maintenance." *Proceedings of the Human Factors Society 28th Annual Meeting* (1984): 913–916.

[13] Harvey, D. "Avco-Pratt—Getting its LHX engine field-ready." *Rotor and Wing International* (February 1987): 26, 28, 56.

[14] Schmitt, J.C. "Design for effective maintenance: safety data provide important directions." *Proceedings of the Human Factors Society 27th Annual Meeting* (1983): 843–847.

[15] Chartrand, P.J. "The impact of organization on labour relations at Air Canada." *Canadian Personnel and Industrial Relations Journal*. vol. 23, no. 1 (1976): 22–26.

[16] Davis, B. "Management and motivation of the aircraft maintenance engineer." *Proceedings of the Air Transport Industry Training Association* (1987) 101–108.

[17] Whitestone, Debra. "People Express." *Harvard Case 9-483-103*. Boston: Harvard Business School (1983).

[18] Smaby, B., C. Meek, J.B. Barnes, and P. Bansai. *Labor-Management Cooperation at Eastern Airlines*. BLMR Contractor's Report, Document 11[8]. Washington, D.C.: U.S. Department of Labor, 1988.

[19] Fotos, C.P. "Continental applies CRM concepts to technical, maintenance corps" and "Training stresses teamwork, self-assessment techniques." *Aviation Week & Space Technology* (August 26, 1991): 32–35.

[20] Stelly Jr., J., and J. Taylor. "Crew coordination concepts for maintenance teams." *Proceedings of the 7th International Symposium on Human Factors in Aircraft Maintenance and Inspection*—"Science, Technology and Management: A Program Review." Washington, D.C.: Federal Aviation Administration, 1992.

[21] Scoble, R. "Recent changes in aircraft maintenance worker relationships." *Proceedings of the 8th FAA Meeting on Human Factors in Aircraft Maintenance and Inspection*. DOT/FAA/AM, Washington, D.C.: Office of Aviation Medicine FAA (1993): 45–59.

[22] Day, S. "Workforce procedures and maintenance productivity at Southwest Airlines." *Proceedings of the 8th Conference on Human Factors Issues in Aircraft Maintenance and Inspection*. Washington, D.C.: Office of Aviation Medicine, FAA (1993): 159–166.

[23] Liddell, F. "Quality assurance at TWA through IAM/FAA maintenance safety committee." *Proceedings of the 8th FAA Meeting on Human Factors in Aircraft Maintenance and Inspection*. DOT/FAA/AM, Washington, D.C.: Office of Aviation Medicine FAA (1993): 167–178.

[24] Taylor, J.C. *Maintenance Resource Management (MRM) in Commercial Aviation: Reducing Errors in Aircraft Maintenance Documentation* (Final Report of 1993–1994 Project Work). Sociotechnical Design Consultants, Box 163, Pacific Palisades, CA 90272, December 1994.

[25] Kania, J. "Panel presentation on airline maintenance human factors." *Proceedings of the 10th FAA Meeting on Human Factors in Aircraft Maintenance and Inspection.* DOT/FAA/AM, Washington, D.C.: Office of Aviation Medicine FAA (1996).

[26] Goglia, J. "Panel presentation on airline maintenance human factors." *Proceedings of the 10th FAA Conference on Human Factors in Maintenance and Inspection.* DOT/FAA/AM, Washington, D.C.: Office of Aviation Medicine FAA (1996).

Chapter 5

Professionalism Now Includes More Communication

OVERVIEW: A New Approach Is All But Inevitable

The professionalism of airplane mechanics has been a major issue for the past decade. In this chapter we'll review what led to this concern and what it means for the future.

The short answer is that professionalism has declined in recent years, for several reasons, and that professionalism can no longer be defined in terms of the ace mechanic, working alone. Professionalism today must include more communication among *all* of the parties working on a given airplane. The mechanics' role must be redefined as an active member of a new organizational unit—the high-communication, multiskilled team.

This strategy functions very well in other industries. Certain industries have been profiting from it for 20 years but have kept it secret for much of that time to preserve their competitive advantage. This strategy pays in so many ways that it's all but inevitable that the aircraft maintenance industry will have to follow the leaders, sooner or later.

This chapter and most of this book focuses on why the professionalism of mechanics is fast becoming the central issue in the organization and management of the aircraft maintenance business. But first, a short review, and then we'll get to the meat.

The cases discussed in this book expose the innards of serious communication errors. As you consider how you'd improve the technical system in these cases—better training, manuals, tools, and so on—don't forget to pay attention to the many social system breakdowns that have eroded mechanics' professionalism and led to recent deadly errors.

Think twice: You'll get a lot farther improving the current social system than you will by tinkering around the edges of the current technical system. It's pretty good, as it is. The social system, on the other hand, isn't. So think about redesigning the socio-technical system to optimize the organization's interface with its turbulent environment. To catch up with a hundred-year relentless pursuit of technical excellence, we must now strive to balance the social system side of the equation.

Think of the social system as the way work is organized and managed to *create* the expectations and performance of the people involved. In essence, the social system is identical with the communication system—it's the structure, the process, and the people involved in answering the question "who talks to whom about what?" The questions of when, where, and how are answered in the very bones of your communication system, in sickness or in health, and the health of communication in aircraft maintenance is not very good.

The social system is one of the biggest problems in maintenance. How do we go about changing it? First, we reprofessionalize the mechanics' work. And then we stop creating more mere mechanical men—no more androids! Changing the structure of the system gives us better communication, coordination, and control. Better worklife satisfaction, too. Not easy, but upcoming chapters show how to get this job done, and done well.

Your communication system today is causing big problems now and it will create even bigger problems in the future if it is not corrected soon.

Big Problem: Some Mechanics Now Just "Factory Hands"

Our earlier review of the Maui accident describes a situation in which hangar mechanics and inspectors were less supervised and received less support than their line maintenance counterparts. This situation can be found in many other North American airlines. It's consistent with the "leave them alone" or *laissez faire* style of management we might expect in a professional setting—a setting in which a professional mechanic is well trained and then certified as competent by the federal government, no less.

But certification alone does not guarantee the best repairs. Consider the newcomers. Leaving newcomers alone with their fresh certificates does not provide the communication needed for the job. What is needed is solid social system guidance and support for technical work. Aloha did not provide such support.

At Aloha, nearly all overhaul work was done at night so the fleet could fly by day. Night work was more acceptable to the younger and the less-experienced mechanics, and they were the ones in the hangar after lights-out in the terminal. But they were also the same employees who most needed additional training, mentoring, and technical information. And they didn't get what they needed.

In short, these less-experienced people did vital work for their company with little social system support. While most of them held A&P licenses, few had much knowledge or experience with modern transport aircraft or modern test and inspection tools. Clearly, this was Aloha management's responsibility, and not that of the mechanics themselves.

With insufficient knowledge and oversight of their own work, these mechanics (those in the hangars and component shops especially) resembled disaffected factory workers more than the able and committed professional mechanics of the past.

Their working conditions and work assignments may have improved, but they had become smaller contributors than their counterparts on the flight line. They were smaller in management's eyes and therefore in their own. Smaller, in fact, in reality.

Encouraged to be *less* independent and *more* like narrow-gauge machines operated by someone else, many of these mechanics experience a gradual shift in their status from multiskilled professionals to something like the stereotypical "low-maintenance" factory hand of the early industrial age—an easily replaced tool to be used as management sees fit.

This big shift amounts to a huge loss in the mechanics' professionalism and in the airlines' margin of safety. Constricted communication practices were both the cause and the result of this drop in the competence and the self-management once expected of aircraft mechanics.

Alarmed at this development, the industry and the federal government worked together to increase the mechanics' professionalism. The proposed new regulation, FAR Part 66, calls for several changes in the mechanics' certificates—most notably the retirement of the Airframe and Powerplant certificates and the introduction of a new one. The Transport Aviation Maintenance Technician (AMT[T]) is a newly defined role that requires considerably more technical training [Ref. 1].

This will help. But other factors have also contributed to the decline in professionalism. They must be dealt with too. Let's look at the history and results of increasing specialization, for example.

Specialization Reduces Professionalism

As specialization increased, communication decreased and along with it professionalism. It was once possible for aviation mechanics to understand the whole airplane. They were highly skilled generalists. All of the information they needed came from the aircraft and the specs. They did not need an elaborate communication system.

As aircraft became more complicated in the 1950s and 1960s, however, the industry needed highly trained specialists to service parts of the plane that required particular expertise. The all-around Airframe & Powerplant (A&P) Mechanic was no longer alone on top. The specialists didn't take over everything, of course, but they did create a new, alternative model for aircraft mechanics. In addition to the traditional A&P generalist, with its focus on the whole plane and its interrelated parts, some mechanics specialized in just one or two of those parts, with the risk of losing sight of the plane as a whole system. And if one A&P's partial repair doesn't fit with another's partial fix, who will solve the problem or even notice it?

But specializing has its appeal. Even among the most broadly trained mechanics, for example, some A&Ps have narrowed their focus to a limited area of special interest instead of trying to stay on top of the whole plane—an increasingly difficult thing to do in any case.

Specialization brought new strains to the workplace. The heyday of the general practitioner was all but gone. Specialization fragmented the work and increased communication needed to keep the work coordinated and safe. In an age of increasing specialization, better ways to organize and manage the work were needed, but no adequate alternatives had yet arrived on the scene. Still, there was no going back to the old ways either.

The invention of the aircraft mechanic specialist came of necessity. The U.S. military led the way in this shift, training fewer all-around generalists and more specialists. Like the commercial airlines, they had no choice. The new planes, together with the diminishing length of military careers, demanded more than the traditional generalist could deliver.

In the late 1960s and early 1970s, the airlines naturally continued their successful longtime practice of hiring mechanics with recent wartime military experience. But the U.S. military had by then trained its Vietnam-era aviation technicians as specialists to deal with the widening gap between individuals' ability to learn and the increasing complexity of modern aircraft. With greater division of labor, the airplanes still got fixed. But professionalism suffered.

At the same time that specialization increased, the larger U.S. aviation scene changed dramatically. For the mechanics, this increased the complexity of their working lives in stressful and unexpected ways. One event after another expelled them from their once-isolated world.

International politics and economic positioning brought about the oil crises of 1973 and 1978. The increasingly free-for-all economy created an atmosphere open to airline deregulation in 1978 and to the fare wars, mergers, and acquisitions to follow. The mechanics could not escape the impact of these events.

Amid the shifting business environment, the aviation industry of the 1980s had to face unhappy changes in its industrial relations scene and stiffer competition from abroad. And it had to deal with the unstoppable revolution in microcomputers and electronics.

A sea change in America's industrial relations climate occurred during the Reagan administration (1980–1988). It weakened organized labor's power to strike, to enroll new members, and to influence politics. Those years included conflicts between the FAA and its air traffic controllers' union, in which 11,500 strikers were replaced with nonunion workers. This opened the way for other employers to hire nonunion replacement workers, resurrecting a little-used 1938 U.S. Supreme Court ruling. The balance of power shifted from the mechanics' end of the pole. But that's not all that changed.

In the late 1970s and the early 1980s, the aging airplanes of the 1960s still looked wonderfully durable and reliable. This good news was coupled with the bad news of unfamiliar cost pressures on the airlines—pressures created by the chaotic rush into deregulated market competition. Since the older planes were holding up so well, it seemed like a good idea to reduce costs by cutting back on maintenance personnel. This time it was the emphasis on the bottom line that changed. And not just for the companies' budgets. Workers took a hit as well.

It was the youngest and least-senior mechanics who were laid off, naturally. Many of them left aircraft maintenance for good. But the good old planes would soon be old planes, period. The older, highly experienced mechanics would soon begin to retire, and the companies would have to hire and train greenhorns all over again. Both the aging aircraft and the more complex new planes coming into the fleet would have to go into the hands of a less-experienced workforce. Even less experienced than those workers laid-off earlier. In hindsight, the vicious cycle is pretty obvious.

The reliable fleet "suddenly" became considerably less reliable. Soon after the layoffs and early rounds of AMT retirements, it became clear that *more* maintenance would be needed, not less. Management had failed to consider the so-called "bathtub curve."

Aircraft designers and engineers can chart the early damage cracks and likely future fatigue cracks of an airplane on a graph that looks like a smiley face, or a bathtub curve, as it's officially called [Ref. 2; 3]. The upturned smile is actually not very happy, just as an overflowing bathtub is not a happy event.

But nobody was watching that. Management was watching current maintenance costs, that's for sure. But they were shortsighted, apparently expecting the future to take care of itself. When the future arrived, it came at the cost of unnecessarily high maintenance bills, high employee training costs and low morale, and greater risks to the flying public.

Movement along the bathtub curve was already well advanced in that fleet of aging, second-generation turbo jets—these were the very reliable, high-altitude jetliners (such as the Boeing 727, 737, and 747 and the Douglas DC9) that were introduced in the 1960s. But a predictable, later-life upswing in the number of cracks came as a big surprise to the airlines. Rudely awakened, they found themselves unable to manage the curve very effectively. Fortunately, aviation people know how to scramble.

After the Aloha Airline B737 fuselage disintegrated in flight, the airlines, manufacturers, and FAA moved quickly to create a series of mandatory inspections and repairs to keep those aging aircraft safely in the air. To get to world-class performance, fast, they've had to closely monitor and upgrade the still-aging fleet every day from day one. More mechanics were needed.

Many new mechanics were certified and hired during 1988–1990 solely to meet the mandated aging-fuselage repair requirements. The class of 1989 by itself nearly equaled the existing A&P labor force of the mid-1980s. The increased maintenance costs were staggering. The risk of errors went up too, in line with the rising number of less-experienced mechanics and the rising number of damage cracks predicted by the bathtub curve. Another correlating curve running upward was the stress it put on people. And there was plenty of stress to go around—on the airplanes, on management, and on the mechanics. Still, that's not all.

Professionalism was being stretched thin by a growing generation gap in the ranks. The class of '89 was younger by 20 years than the existing A&Ps in the companies they joined. To understand how this generation gap affected the work environment, remember that the new mechanics were drawn from an economy with relatively high unemployment. Many were lured by the promise of steady work and big pay after shelling out thousands of their own dollars for the hundreds of hours of schooling required. They came in with high expectations and little fear.

Many of the new hires were formerly automotive sheet-metal workers. Few of the class of '89 were drawn into aviation maintenance by a special love of airplanes. This was opposite the culture of the older, existing workforce. While the older mechanics had not come by their jobs easily, they had grown up in the golden age of flight. They *lived* it, and many of them came from "aviation families."

Another big generational difference between these workers was the nature of the planes they serviced. The airplanes the old-timers had mastered were mechanical. The AMTs had grown very knowledgeable about these planes and they'd grown confident and comfortable with them. Not so with the new workers. Mechanical devices were okay enough, but the youngsters got into the game as products of the modern electronics age. Even if their incoming knowledge of electronics was limited to Nintendo games and laptop computers, they were still way ahead of their elders when it came to learning aircraft electronics.

These differences between the high- and low-seniority mechanics further depressed the already diminished professionalism of the typical AMT. Not surprisingly, the freshmen and the old boys didn't get along very well. The situation called for *better* communication across the generations but *worse* is what the airlines got. The airlines failed to tame the monster, and by and large, they didn't know how.

Professionalism? In this Environment?

It's not easy to be a true professional in the environment we've just described, especially if you're one of the new guys, but the professionalism of the old hands was under duress as well.

The mechanics had to learn to coddle the aging aircraft, *and* they had to learn how to work on the new planes, which were complicated as never before. The third-generation jetliners ushered in microcomputers and other fancy new electronics: the glass cockpit instrumentation, the fly-by-wire flight surface controls, the built-in test equipment (BITE) technology, the ACARS air-to-ground computer communication systems, and the ARINC electronic bus/database. It added up to one thing in the end: Work-related communication problems were barely within tolerable limits, pushing the drop in professionalism toward breaking.

The stress on both generations was enormous. Back in the early 1960s, mechanics had their hands full just trying to comprehend airplanes as whole systems, dynamically integrated single entities of many parts. You'd think that was enough. By the 1980s, however, the mechanics had to deal with specialization, aging aircraft, complex new technology, and the newly emerging business environment. After 1988, all mechanics had new co-workers to cope with. Whether they were from the green or seasoned generation, the environment they were thrown into created an unwelcome strain on the mechanics' already-thin workplace communication system, not to mention their personal communication skills and their tolerance for interpersonal strain.

In the game of football, this would be called piling-on. Against the rules. But aircraft maintenance is not a game. The only certain rule is that any serious communication problem guarantees that you *will* lose ground.

This is worse than depressing, isn't it? It flattens the mechanics' sense of confidence and makes professionalism an empty word. Management, suppliers, and regulators, too, have all had to face up to a long list of new challenges in aviation generally and aircraft maintenance in particular. Greater turbulence has become the norm for everybody in the industry.

Which takes us onto the larger field, the *context* of aircraft maintenance work. The larger system. The overall aviation context, with its interconnected events, has implications thousands of miles and hundreds of hours away. It's a big interactive system. A tiny action in any part of this system can result in a huge and unpredictable reaction anywhere or everywhere in the system.

Mechanics and management need to know this. To operate effectively in today's business environment, we all need to know where we stand in the Big System, the larger context of our work—the department; the location; the division; the company; the industry; the profession; and the general technical, economic, and social worlds of which we are a part. Every part of this large system is both a cause and an effect of every other part. Seeing the world of work this way is an important first step toward effective decision making at every level. The context of the work must be given its due.

Professionalism in Context

Some mechanics have been asking for the big picture—information about the context of their work—and not getting it. Some have never thought to ask. Some in management have been insensitive to the need to share such information. Whatever the cause, the effect of not understanding the full implications of one's work has generated no small amount of unwanted, unnecessary trouble.

Trouble begins when mechanics experience a net loss of professionalism. Although highly trained and conscientious, mechanics' increased specialization has diminished their role as professionals. Doing one aspect of the job well is not the same as doing the whole job well. Professionals do the whole job well, working the whole plane alone if it's a one-person job or working the whole plane as part of an interactive team if it's not. Technicians who have lost contact with the whole job are not professionals any more, but just employees. They are working outside the context of the whole airplane, the whole workforce, and the whole company in its environment.

Management gets pulled down to a lower level of professionalism too, because they now have to make many smaller decisions instead of fewer big decisions. Everybody becomes a de-fanged professional, with their attention drawn to increasingly smaller and more fragmented pieces of the work.

One striking result of specialization is that in just two decades—one generation—many proud, professional, broad-gauge individualists were reduced into narrow-gauge, begrudging job holders, people who have increasingly taken on the character and culture of the alienated factory worker prevalent in American industry during this same period. The headlines called it "the blue-collar blues." No wonder.

For aircraft mechanics as for factory workers, the cause was the same: They were reduced to mere cogs in the machine. However smart, skilled, or socially adaptable, they were still separated from the work as a whole, from their colleagues as working units, and from the larger field that embraced them, sustained them, and all-too-often threatened them. The new mechanics were operating in a far narrower work structure, a structure that made productive workplace communication both more necessary and more difficult.

Unlike the autonomy of the generalist mechanics of yore, the new structure, with its increased specialization and its decreased professionalism, has led management to call for more decision making and coordination from above, instead of from those closest to the work, the mechanics themselves. This is the wrong direction.

As the airplanes and the industry become more complex, management needs *more* intelligent help from the mechanics, not less. The maintenance professional of the future is neither the individual generalist of the past nor the narrow specialist. Specializing leaves a great deal of human potential untapped, and the risk of error is greater than its defenders might suppose. The future will go to the multiskilled, highly communicative team. The team approach is a synthesis: Specialists and others combine forces, providing a highly skilled "team-generalist" perspective to the complex work at hand. The team members collaborate in order to handle the whole better than any one of them could, working alone.

A Parting Thought

For all practical purposes, many AMTs still work alone and in the dark about the relevant larger and even smaller events going on around them. Once they lost their mastery of the whole airplane, they became less than self-directing professionals. Because they no longer had a feel for the whole plane, their group, or the larger context, they naturally began to fit their responsibility to the smaller pieces that they could see. Whoever held the pieces together, it was not them, not any longer, and not even at the level of the single airplane. And if it's not the AMT, who is it?

The risk of human errors and larger system errors naturally increases in such a situation. How to respond? One management scenario said that risk could be reduced only by increasing specialization, backed up with new and extraordinary efforts at management control. Another scenario said that risk could be reduced through technical training, backed up with the normal, cooperative communication that might be expected among the mechanics themselves if their work were structured to make it happen that way.

Such a structure would have to rely on professional-grade communication, commitment, collaboration, and project control among the AMTs themselves. They would be taking on a set of mutually interdependent roles, working together on the whole airplane, and staying in touch with the larger aviation context. There's one good reason to make a change in this direction—today's environment demands it.

Isolated individuals just can't do the job any more. Not without increasing risk and raising costs of every kind. A team of true professionals *can* do the job that is now required. Professionals working together can and do improve quality of work, productivity, costs, and professionalism, all at the same time.

References

[1] FAA, 1996. *Part 66—The New Certification Regulations for Aviation Maintenance Personnel: What They Mean for You.* AC 66-XX (draft, 12/31/96). Washington, D.C.

[2] Marshall, E. "Cracks in geriatric aircraft." *Science* (February 1989): 595–597.

[3] Abelkis, P.R., M.B. Harmon, and D.S. Warren. "Use of durability and damage tolerance concepts in the development of transport aircraft continuing structural integrity program." *Proceedings of the 13th International Committee on Aeronautical Fatigue.* (Also Douglas Aircraft Company Paper 750.5), Pisa, Italy. May 1985.

Chapter 6

The Quality of Communication Determines the Quality of Decision Making

OVERVIEW: Cost Pressures Threaten Communication Quality When It's Needed Most

Containing costs in aircraft maintenance dominates airline maintenance policy and practice. Today these pressures are considerable and are rooted in the market deregulation of 1978 [Ref. 1]. Contract repairs is one way an airline can save on personnel costs during periods of fluctuating work demand.

The independent repair stations, which act as vendors to the airlines, experience these cost pressures even more acutely. These repair stations are largely a phenomenon of deregulation, downsizing, and other reductions in the numbers of mechanics previously employed by the airlines. Many repair stations are start-up organizations, many are small, and they often manage maintenance in the old command-and-control style instead of encouraging and emphasizing professionalism. Although A&P mechanics are employed by some repair stations, these companies also employ unlicensed technicians who are supervised by licensed repairmen. Very few maintenance employees at repair stations are represented by a trade union. These factors keep costs down and in the short run are efficient. But these fundamental differences between the airlines and repair stations, such as the carriers' focus on value and the vendors' focus on costs, can be the source of serious communication problems.

Agreements reached between senior management of the repair stations and their counterparts in the airlines appear satisfactory to both parties. Airlines seem content to contract with repair stations to overhaul aircraft at fixed costs and fixed delivery dates, often with the stipulation that the vendor automatically pays for mistakes and delays that occur when it is clear that they are at fault. But despite that apparent agreement at the top, differences and conflicts occur up and down the separate hierarchies of both vendors and their airline clients. The use of vendors increases the need for effective communication, but makes it less likely to occur.

Here's a scenario for conflict all around. When negotiating the contract between vendor and client, lawyers from both companies meet to work out their differences, typically excluding those in both organizations who are responsible for maintenance—that is, the maintenance managers and the AMTs. Problem number one.

After the work is done, the overhauled or repaired aircraft are delivered and the client gets the bill. But then something wrong is discovered. Who's at fault is not always clear. The airline's maintenance organization says it won't pay for fixing the "vendor's mistake." The vendor pays rather than fight, and then goes looking for an erring employee to punish or discharge. Communication problem number two.

Perhaps the flaw or error is found by diligent and professional AMTs at the airline before it becomes a problem. Maybe, even, the airline's AMTs may go out of their way to find fault with repairs performed by the vendor's largely unlicensed workforce. The airline's management accuses their own maintenance folk of wanting a Cadillac when the company has only paid for a dependable Chevy. Communication problem number three.

For their part, the vendor's mechanics may be reluctant to identify flaws in their own work because their company will have to pay for these flaws, and this doesn't make management happy. Communication problem number four.

These communication problems are all direct results of the built-in split between the two maintenance organizations. They're structural problems.

In Chapter 5, we raised the possibility of safety and quality of work being compromised by mechanics acting (and being managed) like disgruntled factory workers rather than responsible professionals. In this chapter we provide evidence for that assertion and show how decreased professionalism complicates communication all around.

Outside Vendors Increase Risk of Costly Communication Errors

Recent airline accidents involving repairs, inspection, or other work performed at outside vendors has exposed the problems of work quality and communication with nonlicensed mechanics [Ref. 2; 3; 4]. In these cases communication between the airline and vendor was incomplete, and communication between the airlines' mechanics and the vendors' employees was nonexistent. This increases, of course, communication problems, erodes the professionalism of mechanics on both sides of the fence, and increases the risk of errors and accidents.

Ordinary communication problems are difficult enough; they quickly expand with the use of outside vendors and additional lines of communication that may or may not be managed well. Unlicensed repair mechanics may lack the necessary training and the desirable association with other professional mechanics. Because of this they may be less likely to consider the implications of their work than a licensed and professionally motivated AMT. When they work together, communication between unlicensed mechanics and AMTs is often absent or limited, if not tense. It is rarely mutually respectful.

Nonlicensed overhaul shop workers are often treated like factory hands, not AMTs. They are typically told exactly what to do by their supervisors, and they are not expected to deviate or go beyond those instructions. Unless those instructions are perfect and completely understood, the result is a reduction in safe repairs. Look at three recent cases of serious aircraft accidents.

Offshore Engine Overhaul Leads to Serious Injury and Major Property Damage

The first accident took place at dusk on June 8, 1995. It involved the rupture of a compressor disk and the subsequent disintegration of the right side jet engine (a Pratt & Whitney JT8D model) on a DC9 aircraft during its take-off roll [Ref. 2]. The airplane, operated by ValuJet Airlines as Flight 597 from Hartsfield Atlanta International Airport to Miami, traveled some 1500 feet down the runway, caught fire, and burned, having never left the ground. One flight attendant received serious puncture wounds caused by metal fragments thrown from the disintegrating engine. Another flight attendant and five passengers received minor injuries. The airplane's fuselage was destroyed by fire.

That JT8D engine was purchased by ValuJet from a Turkish repair station, Turk Hava Yollari (THY), as part of a package of nine aircraft and five spare engines. THY had last inspected the critical engine parts of that engine some four years prior to the crash. ValuJet installed that engine on the airplane and placed it into service three months before the accident. By the time it disintegrated, the engine had logged a total of 4,433 cycles (takeoffs and landings) since the THY overhaul.

Postcrash examination of the engine's ruptured disk revealed cadmium plating over a one-half inch crack and over severe corrosion pitting. Either the plating had not been removed before crack testing as required by Pratt & Whitney (P&W) specifications, or the corroded and cracked region had not been carefully inspected after the original plating was removed.

Inspection of five other JT8D engines previously owned by THY and sold to ValuJet revealed similar plating over corrosion on the compressor disks in two of those engines. It was beginning to look like a THY standard practice, no matter the risk.

As reported in the NTSB report of the accident, THY maintenance and inspection personnel failed to perform proper inspection of a seventh-stage high-compressor disk, "thus allowing the detectable crack to grow to a length at which the disk ruptured. . ." [Ref. 2 p. 59]. These THY personnel were not operating under U.S. FAA authority or license, but under Turkish authority. Also, there was confusion during the NTSB hearing over whether THY even had JT8D engine overhaul authority during the period from 1986 to 1994 [Ref. 2 p. 47].

Paperwork discrepancies also contributed to the accident. These included an illustration from the P&W repair manual that did not show the 12 holes in the seventh-stage disk where the serious crack occurred. This illustration could have led an inspector to ignore corrosion and cracks around those 12 holes in the part itself.

Another paperwork problem was noted by P&W in a THY workshop review about the time the accident engine was being overhauled. In that report, P&W noted that THY lacked "process sheets" in the Turkish language, which described specific process and repair procedures and that required signatures of the appropriate shop personnel for each significant step in the process.

P&W stated that the job card used at THY, which routed parts through crack detection, did not specify the process, nor did it have a signoff line for completion of crack detection. The NTSB agreed that THY was not using detailed documentation to provide step-by-step guidance of the inspection process [Ref. 2 p. 44]. Since the THY workers were probably operating as factory hands rather than as knowledgeable and responsible AMTs, the inadequate paperwork guidance added to the likelihood of not detecting the crack and thus contributed to the accident.

Remote from one another in space, in time, and in culture, the THY mechanics who inspected the engine and the ValuJet AMTs who installed and maintained it did not—could not—coordinate their work or provide professional respect, support, and motivation for one another.

The Atlanta accident is not just a case of technical problems; it's a case in which a poor system of communication led to technical problems that destroyed an airplane and injured seven people.

Factory Worker's Propeller Repairs Lead to Serious Accident

In the second accident, a Brasilia Embraer aircraft, the Atlantic Southeast Airlines Flight 529 on August 21, 1995 (from Atlanta's Hartsfield International Airport to Gulfport, Mississippi) lost a propeller blade from its left side engine while climbing through 18,000 feet. The pilot attempted to return to Atlanta, but was unable to do so, eventually crashing near the municipal airport at Carrollton, Georgia. Of the 29 people on board, one crew member and eight passengers died; 12 sustained serious injuries, while 8 more received minor injuries.

Following its investigation and hearings, the NTSB concluded that the cause of the accident was a fatigue crack that originated from multiple corrosion pits in the propeller's hollow core (or "taper bore") [Ref. 3]. That crack was later determined to be already about 3/32 of an inch long (0.070 in.) when it was initially detected during a federally mandated Airworthiness Directive (AD) on-wing inspection 28 months before the accident. The inspector who found the crack was a licensed inspector employed by a vendor under contract to Atlantic Southeast Airlines.

Less than one month after the AD inspection, the suspect propeller was removed from the engine and sent to the manufacturer (Hamilton Standard Company). The manufacturer further inspected, repaired, and released the blade for service, without the crack ever having been detected. That second inspection, repair work, and signoff were performed by an employee of the manufacturer—an unlicensed "technician" (a grade of mechanic below "repairman") who had less than 200 hours experience in that type of repair work.

After undergoing further repair work at the manufacturer, the propeller was returned to the airline in August 1994, exempted from further routine inspections for cracks in the taper bore, and installed on the accident aircraft a month afterward. Less than one year later, the propeller disintegrated at the very point identified by the original inspector during the on-wing AD inspection. The AD inspector who originally detected the crack had not been apprised of the manufacturer's action, nor had he been consulted farther down the line.

The proper technical procedure for detecting cracks was misapplied by the Hamilton Standard technician. The NTSB concluded that the misapplication was also the result of poor communication practices within Hamilton Standard. The decision to modify the technical procedure had been noted among the manufacturer's top engineering managers, but not filed with the FAA, the company's FAA Designated Engineering Representative (DER), or the AD inspection vendor's personnel. The manufacturer's procedure was documented thereafter in a memo that didn't mention special circumstances and conditions. The substance of the memo was then orally communicated by the Hamilton Standard factory manager to his staff, including the technician in question who, by his misapplication, showed that he had either misunderstood the procedure or it had been misstated to him [Ref. 3 p. 55].

The airline Maintenance Department cannot manage workplace communication for its vendors. And the vendor's factory hand does what he is told to do, usually without the expectation that further thought or decision making may be necessary.

The string of communication problems in the Carrollton, Georgia, crash can be attributed to cross-boundary difficulty and to the lack of a professional-grade communication system. The technician was not solely responsible for those deaths and injuries.

The Deadly Effects of Mislabeled Cargo

On May 11, 1996, a DC9-32 aircraft, ValuJet Airlines Flight 592 (from Miami International Airport to Atlanta's Hartsfield Airport) caught fire shortly before takeoff. Seven minutes after takeoff, the pilots noticed the problem, and five minutes later the plane crashed into the Florida Everglades at about 450 miles per hour. The pilot apparently attempted to return to the airport but was unable to do so.

All 110 aboard were killed.

The NTSB concluded that the fire leading to the crash was caused by the combustion of 144 oxygen generator canisters loaded in the aircraft's cargo hold [Ref. 4].

From testimony at the accident investigation, it was determined that during March 1996, employees of a repair station contractor, SabreTech Aviation, removed those canisters from two other ValuJet MD-80 aircraft and did not place safety caps on them. In addition, the mechanics labeled the removed canisters with tags that marked them as reusable, but not necessarily empty (most of them were not). After the job was completed, a SabreTech supervisor signed the final block. Despite prior discrepancies, that supervising licensed mechanic also signed the block confirming that the mechanic blocks had all been signed off, even though the work cards specified that safety caps *must* be installed on oxygen generators when they are removed from an aircraft [Ref. 4 p. 113]. The supervisor's signoff block did not require that the work be inspected. Those errors were compounded by not labeling the charged canisters as hazardous material, thereby allowing them to be carried on a passenger aircraft.

In early May, some two months after work on the two MD-80s began, a SabreTech stock clerk who was repacking the generators noted their tags and assumed (incorrectly) that they were empty. He then sealed the five cartons and mislabeled them as containing empty (therefore inert) oxygen canisters.

By this time the contract period for completing work on the two MD-80s was expiring, and SabreTech was about to lose $2500 per aircraft for every day the work remained incomplete. Avoiding additional losses must have created considerable pressure throughout SabreTech. It was known, for example, that SabreTech management required all maintenance personnel to work without days off until the MD-80s were delivered [Ref. 4 pp. 13–16].

An unlicensed SabreTech shipping clerk—who was directed by SabreTech management to send the canisters back to ValuJet—had consequently created the shipping ticket that labeled the five cartons incorrectly. A driver subsequently drew the mislabeled boxes containing the 144 canisters from storage and left them on the ramp, where they were loaded in the hold of Flight 592. The SabreTech employee doing the loading recalled later that the contents of the cartons had "clinked" as he set them, unsecured, on top of spare aircraft wheels and tires in the front of the DC9's hold.

It is assumed that one or more canisters subsequently tumbled out of the container and detonated its built-in percussion cap, creating an oxygen-generating chain reaction that produced great heat, igniting the wheels and tires beneath the canisters and providing an oxygen-rich environment fueling the flames even more.

Among other findings of inadequate oversight and supervision, the NTSB found that ValuJet's maintenance management had neither clearly warned their contractor (SabreTech) or the contractor's employees nor provided them with the appropriate material (e.g., canister safety caps) for the canisters' safe handling [Ref. 4 pp. 123, 136–7]. SabreTech's mechanics did not exercise good judgment in their handling of the canisters or in the accompanying documentation. They did not think ahead to possible consequences of their acts, each of which was a minor infraction by itself. They were just taking orders and *almost* following the work card's instructions. Under pressure to finish the contracted work, SabreTech management made less-than-optimal decisions to pack and ship the canisters—decisions they might have questioned under normal circumstances.

The NTSB listed, among other causes, the following two:

1. The failure of SabreTech to properly prepare, package, identify, and track unexpended chemical oxygen generators to ValuJet for carriage.
2. The failure of ValuJet to properly oversee its contract maintenance programs to ensure compliance with maintenance, maintenance training, and hazardous-materials requirements.

Employee turnover rates at SabreTech were as high as 50% at the time of the accident. Many of those employed were independent subcontractors employed on a casual basis. Seventy-two technicians working on three shifts in the contractor's Miami hangar logged 910 hours to the job of replacing the oxygen generators on the ValuJet MD-80s. Those 72 technicians were supervised by one A&P-licensed SabreTech employee and by two other subcontractors who were not SabreTech employees.

Commenting on this communication and management challenge, the NTSB accident report states that "SabreTech followed no consistent procedures for briefing incoming employees at the beginning of a new shift and had no system for tracking which specific tasks were performed during each shift" [Ref. 4 p. 15]. Thus, the technicians who removed and mislabeled the canisters may never have seen one another, or those who later loaded the canisters into the accident aircraft, and they may not even have met their SabreTech supervisor on a regular or frequent basis. And those SabreTech employees were not in communication with ValuJet's AMTs. Without a carefully designed communication system in place, the vendor system compromises both communication and professionalism.

Why hadn't both SabreTech and ValuJet management communicated the dangers of hazardous materials more clearly? Complacency, inattention, or poor judgment? An editorial in *Aviation Week* for September 1, 1997, noted that although SabreTech admitted responsibility for at least some errors leading to the crash of Flight 592, ValuJet had not done so [Ref. 5]. In fact,

ValuJet attempted to lay the entire blame on their vendor. Reports state that ValuJet management claimed that SabreTech had "deliberately and deceptively" mislabeled and shipped the canisters in violation of Federal Aviation Regulations as well as its own procedures. The company claimed that SabreTech "deliberately obscured the shipment" of oxygen canisters to get the potentially hazardous material off the premises at Miami Airport prior to an inspection by a new airline customer [Ref. 6]. The September editorial offers specific advice to ValuJet in expecting the airline's own supervisors and inspectors to double-check contractors' work, and to do so around the clock if that work is conducted over three shifts [Ref. 5]. ValuJet, the editorial notes, did none of those things.

Missing O-Rings Lead to In-Flight Shutdowns and Nearly Cause Disaster

Unfortunately, the case of Flight 592 is not the first, or only, instance where an airline didn't provide enough information to guarantee satisfactory understanding and coordination of the work. Airlines may have the hardware, the bulletins, the manuals, and the technical skills, but they don't have the world's best workplace communication systems to tie it all together. Not by a long shot.

What follows is another clearcut case of poor maintenance communication. On May 5, 1983, at 9:15 a.m., some 20 minutes after Eastern Airlines (EAL) Flight 855 departed Miami International en route to Nassau, Bahamas, a low-oil indicator on the No. 2 engine caused the pilot to shut down that engine and turn back to Miami. On that return flight, the aircraft, a Lockheed L-1011, experienced the flame-out of its other two engines. The pilots were able to restart No. 2 engine long enough to land safely at Miami, but the situation was so serious that the NTSB officially termed the event a true "accident" [Ref. 7].

Technically, the cause was simple and easily identified—chip detector plugs had been replaced in the engines without installing new O-rings, and the oil had simply run out around the plugs. During the preceding two years, Eastern pilots reported twelve previous instances of L-1011 oil loss due to plug and seal leaks. Eight of those involved in-flight shutdowns and seven of the eight resulted in unscheduled landings. Twelve serious incidents did not effect a correction in time to prevent the near-death experience of Flight 855. A fairly small technical problem remained unsolved for so long because the communication system failed to get the job done.

Company management had acknowledged the problem early on, and they had ordered changes in job cards to correct the problem. In addition, training bulletins were issued to broadcast the problem and the changes. The company required that such bulletins be posted in the maintenance area, and it required all foremen, leads, and mechanics to read them. Why did these two communication initiatives fail? Complacency—a communication risk that had not been recognized and addressed.

The period of 1980–1983 was marked by difficult industrial relations among the company's management, the mechanics' union, and the pilots' and flight attendants' unions. In fact, hardly a month before Flight 855's accident, a strike by mechanics had been narrowly avoided, but the resulting contract led management to announce the elimination of 1600 jobs (including 1200 mechanics) [Ref. 8 p. 8]. Despite this added stress, there was never any suspicion that prior labor management animosity clouded the mechanics' judgment. The NTSB simply noted that the "routineness of [the chip detector] task [itself] can lead to unquestioning, nonattentive action" (complacency).

Despite widespread management awareness of the chip detector problem and the "fixes," the mechanics involved in the May 5th accident were not aware of the proper procedures. They testified that they had not even read the training bulletin describing the proper procedure to replace chip detectors.

During the investigation, it was discovered that these mechanics had never replaced the O-ring seals on chip detectors [Ref. 7 p. 31], even though they had performed over 100 detector changes [Ref. 7 p. 23]. During the NTSB hearings, the airline's management speculated that, in the past, O-rings had been replaced by people other than the mechanics doing the work on Flight 855's aircraft, or that recycled chip detector plugs had been reissued to mechanics with the old O-rings still installed. On the night the plugs were replaced on Flight 855's aircraft, there were no chip detector plugs in the cabinet where the Miami foreman usually kept them. The mechanics assigned to the work merely installed the new plugs they obtained from stores, without O-rings. Why didn't they know the correct procedures? Why hadn't they read and/or understood either the training bulletin or the job card? Why hadn't management been able to communicate the procedure orally, in writing, or by demonstrating it? Why had the mechanics acted with complacency?

The "missing O-ring" case is, of course, about insufficient attention to business. Despite distractions, managers and mechanics must strive to communicate vital work-related information clearly, at all times, even when there are labor problems. This case shows that airline mechanics are sometimes provided a simple task list (like a job card) accompanied by contradictory technical information. If mechanics perform these tasks without attention or thought, serious errors may follow. Even the simplest technical task requires critical thinking, a process that thrives in an environment of open, accurate, and complete communication among the people involved. Attentiveness is required of management and mechanics alike. Only an excellent multimedia communication system can create and sustain attention to detail and to the larger picture at the same time.

At present, this ideal workplace communication environment exists nowhere in aviation maintenance. Some efforts intended to ensure better communication do exist, but they are few and are not completely developed. In some cases management's attempt at *two-way* communication simply requires AMTs to sign or initial a roster indicating they have actually seen the document. This is *not* two-way communication. As with EAL's O-ring bulletins, it does not succeed as one-way communication either.

Without "face time," without live-action questions and answers, demonstrations, and discussion, there's no telling whether the information was received, what exactly management meant by its bulletin, what the AMTs understood, or what special arrangements each group might require of the other in order to get the job done. The bulletin has its uses, but by itself it is about as minimal as workplace communication can get. Job cards can also be unread or incorrectly understood.

Relying on bulletins and job cards alone is inexpensive but not very effective, especially if it puts the airplane and its passengers at greater risk. Nevertheless, such minimal written communication is the medium of choice in the industry. Important technical matters are communicated almost entirely through printed bulletins, written job cards, and manuals; this is standard

The Quality of Communication Determines the Quality of Decision Making

practice in aviation maintenance. It's a practice that contributed to the O-ring accident on EAL's Flight 855 and the dozen incidents that preceded it.

Eastern was not alone in this kind of dangerous communication gap. Only a few airlines reported similar problems with master chip detectors prior to the EAL accident [Ref. 7], but they all have their own version of such communication problems in other technical matters—problems that have not as yet brought on an NTSB investigation.

The investigation of the EAL accident spread blame on all maintenance parties, and not just the management that failed to ensure that its communication hit the mark. The NTSB singled out the mechanics for special mention: The mechanics "failed to perform their duties with the *professional* care expected of an A&P mechanic" [Ref. 6 p. 43, emphasis added].

There's that word again: professional. In this case, the word *professional* is clearly meant to include the expectation that the mechanics themselves, and not just the boss, accept responsibility for keeping current, alert, and competent in their work. And professional also means that the safety implications of all maintenance tasks are understood, recognized, and committed to.

Even if the technical bulletin continues to be the only "communication" offered, professional mechanics will study it thoroughly and as soon as it's available. They will ask questions. And they will check their understanding against that of other AMTs and engineers or technical managers. Just to be sure. Just to be safe.

That's how professionals manage their personal workplace communications, with or without further help from management. Professionals know that the quality of their decisions depends on the quality of their communication. Just as they leave nothing to chance with technical tasks, neither do they leave anything to chance in professional communication.

That's more or less what the NTSB had in mind when it drew attention to lack of professionalism in those aircraft mechanics at EAL. Across the industry, numerous AMTs are clearly professionals, while others are less sure. This can be said of managers as well. The quality of their communication is a direct reflection of their own professionalism.

A Parting Thought

Airline maintenance managers must continue supporting and increasing their own professionalism and that of their AMTs. The details of these cases provide instructive examples for acting to prevent similar problems elsewhere. But the specifics are not as important as the pattern. A minimal program of written technical documents is not sufficient for company internal communication. Effective communication with outside maintenance vendors is even harder to achieve. Both should be at the top of your agenda for corrective actions.

References

[1] OTA (Office of Technology Assessment), *Safe Skies for Tomorrow: Aviation Safety in a Competitive Environment*. OTA-SET-381. Congress of the United States, Washington, D.C.: U.S. Government Printing Office, 1988.

[2] National Transportation Safety Board (NTSB), *Aircraft Accident Report: Uncontained Engine Failure/Fire Valujet Airlines Flight 597, Douglas DC-9-32, N908VJ, Atlanta, GA, June 8, 1995*. NTSB/AAR-96/03. Washington, D.C., 1996.

[3] NTSB, *Aircraft Accident Report: In-flight Loss of Propeller Blade Forced Landing and Collision with Terrain, Atlantic Southeast Airlines, Inc., Flight 529 Embraer EMB-120RT, N256AS, Carrollton, GA, August 21, 1995*. NTSB/AAR-96/06. Washington, D.C., 1996.

[4] NTSB, *Aircraft Accident Report: In-flight Fire and Impact with Terrain, Valujet Airlines Flight 592, Douglas DC-9-32, N904VJ, Everglades, Near Miami FL, May 11, 1996*. NTSB/AAR-97/06. Washington, D.C., 1997.

[5] "Playing hookey for lessons of Flight 592." *Aviation Week* (September 1, 1997).

[6] McGraw-Hill News Services, "ValuJet asks NTSB to delay Flight 592 report." *Aviation Daily On-line Services* (August 14, 1997).

[7] NTSB, *Aircraft Accident Report: Eastern Air Lines, Inc., Lockheed L-1011, N334EA, Miami International Airport, May 5, 1983*. NTSB/AAR-84/04. Washington, D.C., 1984.

[8] Smaby, B.; C. Meek, C. Barnes, J. Blasi, and P. Bansai. "Labor-management cooperation at Eastern Air Lines." *Bureau of Labor-Management Relations and Cooperative Programs Report BLMR 118*. Washington, D.C.: U.S. Department of Labor, 1988.

Chapter 7

Up Your Professionalism

OVERVIEW: Today's Mechanics Must Rise to New Levels

Various reforms are underway to increase mechanics' technical skills and to enhance the field as an attractive career choice. Job requirements have been raised, the job has been redefined to include new levels of proficiency, and each of these has acquired a new name to reflect the level of professional skill expected. It's time to take a closer look at what professionalism means today.

Let's get specific. We've been using the term "AMT" broadly up to now. Actually, it's the new name for top-grade mechanics only. It's shorthand for "Transport Aviation Maintenance Technician" (officially "AMT-T," [Ref. 1]). It's the Ph.D. of maintenance. You can't get any higher than that.

The term has been used in its loose sense up to now to include QC inspectors and certificated Avionics and Electronics technicians, as well as the top mechanics. But the AMT-T designation is meant for only those who have passed a rigorous new training program and qualifying tests for the new-style AMT-T certificate. Others may qualify too, if they already hold the old certificates and they're already doing the work of an AMT-T.

Getting the nomenclature straight is an adjustment still under way. Not everyone who touches the plane is an AMT. The term is more restrictive now, and it's used only in the United States. In the U.K., Canada, and other Commonwealth countries, the title of the top job is still "Aircraft Maintenance Engineer" (AME) or some other combination that includes "engineer" in the traditional sense of a skilled machinist or mechanic. In those countries, the use of the term AME does not refer to the degreed electrical, mechanical, or aeronautical engineers who sit at desks with computers. Aviation mechanics everywhere are the people who see, hear, touch, and smell the airplanes all day every day. Their tools are micrometers, wrenches, oil guns, and a host of sophisticated testing equipment. They're in the business of applied engineering. They're *practical* engineers, by any title.

At this writing, the new AMT designation is not yet official. It's recommended for the Federal Aviation Regulations standard (FAR Part 66) for the proposed advanced certification [Ref. 1]. Over the course of the past five years, however, the informal title changeover that started before 1984 [Ref. 2] has come into widespread use without waiting for the government's stamp of approval.

Now that's the short of it. The long of it amounts to the story of how the new AMT title and several others came to pass. These new titles are not somebody's whim. These new certificates were created to increase the professional identity and proficiency of airline mechanics. A need for this became apparent in the post-1988 period as thousands of new mechanics were hired to modify aging airplanes. The influx of new hires in the late 1980s coincided with powerful changes in politics and economics, and in technology and society as well. These changes taken in the context of flight safety put enormous pressure on the airlines and their people.

During this time, while the new mechanics were tossed around in the crosswinds of change, the process of deprofessionalizing-by-way-of-specializing accelerated. The new mechanics came to see themselves as insignificant extras—-mere "grease monkeys," "wrench twisters," or even "just mechanics" [Ref. 3]. Add to that unhappy development the already-eroding professionalism of their senior colleagues, and the picture was not pretty. The good news is, it is now finally being corrected.

The following description of professionalism is presented during "Human Factors in Maintenance" training sessions for mechanics in a large U.S. airline (see Table 7.1). The participants uniformly confirmed this description and endorsed its new points of emphasis in their training. There were also those who wondered aloud why they didn't receive higher wages if their preparation compared so favorably with pilots and degreed engineers. But laying the issue of relative pay aside, they are not just grease monkeys—they're professionals.

TABLE 7.1 GREASE MONKEY OR PROFESSIONAL, YOU BE THE JUDGE

The airplane sat on the bitter cold ramp with the cowling open and a lone figure hunched over changing an ignitor box. The trip was late, everyone wanted to get on their way. Several people vented their frustration about being delayed; they blamed the mechanic. The mechanic continued until the job was completed per the maintenance manual. Then signed for his work. Let's look at just three things about a mechanic for just a minute. The A&P mechanic is the only maintenance professional certificated by the federal government. The school must be an FAA-approved Part 147 Aviation Maintenance Technician School. The school must cover 43 subject areas and provide a minimum of 1900 hours of instruction. To take the tests based on practical experience, the applicant must show the FAA a minimum of 4800 hours of airframe and powerplant experience. A good way to remember this is to compare them against the 1500 hours for an Air Transport Rating or 1860 hours for a 127 college credit B.S. degree. The average startup tool box will cost the mechanic approximately $3000–$4000. Quietly through the night, mechanics throughout the world work to ensure the safety of thousands of passengers every day. The mechanic is the muscle and bones of aviation. Can you picture aviation without mechanics?

Source: Adapted from O'Brien [Ref. 3]

Professional AMTs Need Better Communication Than Mere Grease Monkeys Do

Not all mechanics want more communication of any kind, but many do, specially the younger ones [Ref. 4; 5]. The new AMT workforce asks for more. It is important to communicate the company's core purpose to mechanics, inspectors, planners, and their supervisors. But the core purpose is not always as clear as some might like to believe. Nor is it emphasized in a way that boosts the mechanics' professional desire to know and to pursue the company's main business as their own main business. Management often does not communicate the single issue of most importance: purpose. The point we raise here is not one of quantity of communication, but of its quality.

Transmitting information does not ensure that it will achieve its intended purpose. All too often we transmit without a clear purpose. The barrage of bulletins, logbooks, and AD notes could better serve their purpose if they were part of a larger effort that includes face-to-face discussions, hands-on demonstrations, and the like. When these kinds of things are seen as important to building a clear and shared purpose, then the paper flood can be understood as back-up reference material—needed but not in itself the guarantor of purpose-driven decision-making on the job. Purpose-driven decision making begins with a clear understanding of organizational purpose.

What is most important is the company's *unique* purpose, distinct from that of other airlines or from the industry as a whole. Mechanics must see how their own life/work purposes are related to the company purpose. To help the mechanics meet their own purposes along with that of the company's, the so-called "big picture" must be communicated thoughtfully and carefully. There are several reasons for this:

1. New employees in airline maintenance no longer come from the military in the numbers they once did. In the past, ex-military mechanics were accustomed to command-and-control organizations where explanations and reasons played little part.
2. Military training itself has changed. Today's enlisted personnel are encouraged to be involved and are provided with information and explanations, as well as direction [Ref. 6].
3. Today employees in U.S. industries expect greater involvement and participation than did their predecessors. Many new airline mechanics first gain experience in other industries where more open communication is encouraged. Their families, schools, and churches generally encourage more open communication and direct involvement in decision making than in the past.
4. The causes or logic behind work processes are no longer apparent at first glance. Sometimes the only way to understand is to ask, to research, or to be told "why"—and then to listen, ask more questions, and keep it up until it all falls into place. Beats airplanes falling all over the place!

In the interests of everyone, the airlines can do more to help maintenance people rise to the level of true professionals, in their individual work and as part of a team of professionals tackling the whole plane. For their part, mechanics can also step up to the challenge in a bigger way.

Professionals don't wait for everything to be handed to them on a silver platter. These are dirty-knuckle working pros who want to do the job better and better. They don't need coddling. They need blunt, complete, on-the-job, work-related communication. That means communication about the point of it all above everything else. Purpose precedes practice.

There's no alternative to this kind of improvement. In turbulent times like these, only the integrated efforts of the best people will do. Only the high-communication professional *team* can handle the growing complexity. Professionals, not shop hands. Communication must attain new levels of excellence for mechanics to rise to their highest standards.

Professionalism Defined and Delivered: The 4 Cs

What's the difference between professionals and nonprofessionals? Traditional professions (physicians, lawyers, engineers, teachers, and public accountants) all share common characteristics (see Table 7.2). Key are a high degree of competence in their field (usually from a stringently defined educational process and examination); the authority and ability to make decisions affecting the control of their work; a commitment to the greater public good provided by their work; and being central in importance to their work.

TABLE 7.2 CHARACTERISTICS OF PROFESSIONALS: THE 4 Cs

- Competence and skill
- Control of own work
- Commitment to higher purpose
- Central to the work process and product

Source: Adapted from Taylor and Christensen [Ref. 7]

It has been shown elsewhere [Ref. 8] that these four characteristics are integral to other occupations also, including aircraft maintenance technicians. The proposed new AMT-T training curriculum and examination are designed to ensure high-level competence. AMTs continue to be responsible for approving aircraft for return to service, and for the high degree of control they have over their work. The proposed FAR Part 66 recognizes the central role AMTs play in ensuring aviation safety [Ref. 1 p. 3].

Only one element of professionalism remains partly unaddressed in the proposed new AMT certificate—the mechanics' *commitment* to the overriding goal of air travel safety. Most mechanics already embrace this commitment; they bring it to the job on their own. But understand that such a professional commitment remains unacknowledged in the new regulation because commitment cannot be regulated. This commitment is a natural result of the other 3 Cs, the integrity of the mechanics themselves, and the organization's culture.

New Classifications Mean More Communication Skills

Federal Aviation Regulation Part 66 proposes changes to several classes of maintenance certificates, among which are the A&P "mechanic" and the aviation "repairman." As noted previously for airline mechanics, the A&P certificate will be replaced by the new AMT-T certificate. There have been a couple of changes in the traditional repairman certificate too.

In the past, a repairman license was created by the FAA in the name of an individual mechanic, but it was issued to the *employer's* repair station. In order to qualify for the certificate, the individuals were required to have at least 18 months of experience on the task specified for certification and have completed specialized training as well.

Because the repairman certificate was officially retained by the aviation maintenance organization, certified repairmen exercised the privileges and responsibilities of their certificate only while employed by that company. When repairmen quit or got fired from a repair station, they could not take the certificate with them, nor could the employer use it again. The proposed FAR Part 66 will change this.

The proposed "Aviation Repair Specialist" (ARS) certificate will be issued for additional and new specialties such as nondestructive inspection (NDI), composite structure repair, metal structure repair, and aircraft electronics. The changes to the proposed ARS certificate will also include portability—the ARS certificate thus stays with individuals if they leave a repair station's employ, but they cannot use it until they are employed again by another repair station.

More Professionalism or Less?

The impact of the new Part 66 proposal should be considerable. On the positive side, the training improvements in the new ARS certificate should improve professional *competence*. Also, the portability of the new ARS certificate should positively affect repair technicians' sense of *control* and their professional *commitment* to air safety. Beyond the changes in the proposed FAR Part 66, true professionalism requires a fundamental change in the basic assumptions of some repairmen and their managers. The continued use of traditional management practices can continue to limit the *control* ARSs have in making technical decisions.

The growth of outsourcing can impact professionalism negatively, as we have said before. And outsourcing is increasing, not a positive development for enhancing professionalism. Although a sizable proportion of vendor mechanics hold A&P certificates, most others are merely employees who report to the certificated repairmen.

The expanding numbers of new repairmen and uncertificated technicians raises the need for quality workplace communication. These lower-skilled mechanics require more communication not less, and they're not getting it.

The FAA tracks and reports the total number of valid mechanic certificates every year [Ref. 9]. This number represents all new certificates issued during the year minus all certificates held by persons over 75 years old. This annual total thus does not reflect the number of people actually working as aircraft mechanics. This number is lower. For instance, the 1990 U.S. Census revealed that only 166,486 workers reported their occupation as "Aircraft Mechanic" [Ref. 10], which is less than half of 344,282 certificates reported by the FAA for 1990 [Ref. 9]. Figure 7.1 plots the number of certificates held between 1988 and 1996, as well as the 1990 census number.

FAA statistics that count repairman certificates in a category apart from A&P certificate holders have only become available since 1995. Prior to that, holders of repairman certificates were included in the numbers for the total A&P mechanics category. This improvement in reporting gives us an accurate idea of the ratio of working A&P mechanics to repairmen. Most people holding repairman certificates are probably currently employed in aircraft maintenance. Thus, their relative proportion in the FAA annual statistics is less of an overestimate than the numbers for A&P license, which represents license holders irrespective to their current work status or place of employment.

The FAA statistics for 1995 and 1996 [Ref. 9; 11] indicate that repair specialists (numbering 61,233 and 50,768, respectively) represent 1 out of every 8 or about 12% of all maintenance certificate holders (totaling 466,517, and 430,775 for those years). That is a sizable proportion, but as noted above, it is

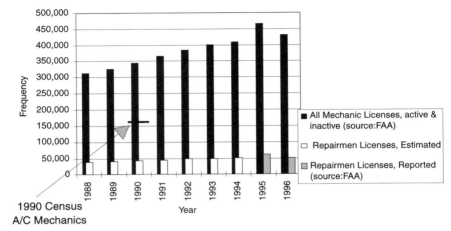

Source: FAA Aviation System Indicators [Ref. 9]; FAA Administrator's Fact Book [Ref. 11]; Census [Ref. 10]

Fig. 7.1 Number of mechanics certificates held, 1988–1996.

probably an underestimate of the true proportion of repair specialists actively employed. However, using two conservative assumptions we get a number that is probably more accurate.

The first assumption is that in 1990 active repairman licenses accounted for about 12% of the FAA numbers, just as they did in 1995 and 1996. The second assumption is that the 1990 U.S. census number of 166,486 is a reasonable estimate of the total aircraft mechanics (with and without certificates) employed in the United States at that time.

Following the first assumption, 12% of the 344,282 total 1990 FAA certificate holders equals 41,314 active repairmen licenses. Figure 7.2 graphs aircraft mechanics employed in 1990.

Using the second assumption, 41,314 licensed repairmen is 25% of the 1990 census total of 166,486; this is a significant proportion of the *active* mechanic workforce. In addition, we can only wonder how many of the remaining 123,623 aircraft mechanics were the uncertificated employees supervised by A&Ps and repairmen. Fig. 7.2 also shows those calculations.

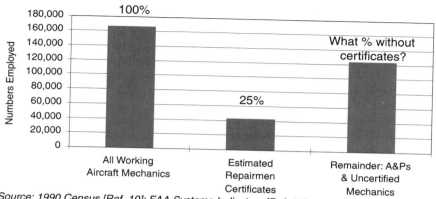

Fig. 7.2 *Aircraft mechanics employed in the United States in 1990.*

With the current trend of outsourcing maintenance repair work, the proportion of repair specialists and unlicensed technicians in the aviation maintenance population seems likely to increase. What will this do to the professionalism of aviation maintenance? What kind of communication and organization culture will result? What new problems? Nobody knows for sure.

Some Parting Thoughts

What we do know is that the increasing numbers of new repairmen, in-house and outsourced, will require an increase in professional-grade communication in both houses. Think about it:

1. Technical *competence* can be maintained and increased only with more training in the classroom and on the job. Training is communication.
2. *Control* and *centrality* can be increased by the new regulations for signoff authority and certificate portability, already in the works, and by better workforce organization to augment the voices of the AMTs and ARSs in quality control and safety assurance. Currently, this is the industry's largest untapped potential. Organizational improvement is another term for communication system improvement.
3. *Commitment,* along with the other 3 Cs—centrality, competence, and control—is at greater risk because of the increasing number of lesser-skilled mechanics at the outside repair stations and by the difficulties of

communicating effectively across the boundary that separates the airline maintenance crews from their outside vendors.

More schooling is needed, for sure, but that's not all. Technical training alone is not enough. In a recent interview [Ref. 12], the director of one of America's oldest and largest aviation training organizations made it clear that "it takes more than just technical information to be a professional in aviation maintenance. It takes communication."

References

[1] FAA. "Part 66—The new certification regulations for aviation maintenance personnel: What they mean for you." AC 66-XX (Draft, 12/31/96).

[2] Strauch, B., and C.E. Sandler. "Human factors considerations in aviation maintenance." *Proceedings of the 28th Annual Meeting of the Human Factors Society.* (1984): 913–916.

[3] O'Brien, Bill. "Grease monkey." *Aircraft Maintenance Technology* (September 1986): 64–68.

[4] Forth, K.D., "Squeaky wheel gets grease—and training." *Aviation Equipment Maintenance* (August 1993): 56.

[5] Taylor, J.C. "Maintenance organization." In W. Shepherd, et al., *Human Factors in Aviation Maintenance Phase 1: Progress Report.* Washington, D.C.: Federal Aviation Administration, Office of Aviation Medicine (1991).

[6] Rogers, A.G. "Organizational factors in the enhancement of aviation maintenance." *Proceedings of the Eighth FAA Meeting on Human Factors in Aircraft Maintenance and Inspection.* DOT/FAA/AM, Washington, D.C.: Office of Aviation Medicine (1991): 43–63.

[7] Taylor, J.C., and T.D. Christensen. *STS Handbook.* South Bend, IN: STS Publishing 1991, pp. 53–54.

[8] Taylor, J.C., and D.F. Felten. *Performance by Design: Sociotechnical Systems in North America.* Englewood Cliffs, NJ: Prentice Hall, 1993.

[9] FAA. *Aviation System Indicators—1996 Annual Report.* Washington, D.C.: U.S. Department of Transportation, 1997.

[10] U.S. Census Report. "Equal employment opportunity for United States," *1990 Census.*

[11] FAA. *Administrator's Fact Book.* Washington, D.C.: U.S. Department of Transportation, FAA/ABC-100 (May 1997).

[12] Jackman, F. "Interview with Mike Lee, flight safety international." *Overhaul & Maintenance* (Jan.–Feb. 1997): 12–14.

Chapter 8

Mechanics Cannot Thrive on the Written Word Alone

OVERVIEW: Effective Communication Is Multimedia

We are by definition multimedia creatures. We have multiple senses with which we process information, and we all process information differently. One person picks up visual cues more readily, while another is better with audio cues. Thus, a solid foundation of written communication must be used in tandem with other communication addressing the same point.

We have called the main point "purpose."

Think about McDonald's. Like many companies, McDonald's succeeds as well as it does in part because it uses a "high-cost" multimedia approach to achieve its purpose: to get our attention, our loyalty, and our bucks. We consumers are the target, and McDonald's managers and employees are in the middle of that target—they are the true bull's eye of McDonald's multimedia approach.

If McDonald's employees don't aim at a well-defined customer target, it will be because their managers have first failed to "shoot straight" with them— that is, lay out the company's official purpose, in writing. We recently walked into a neighborhood McDonald's for a coffee and sandwich. While there we thought we'd learn more about McDonald's stated philosophy: QSCV (Quality, Service, Cleanliness, and Value). We asked the young man at the counter about QSCV. He said, "I'm supposed to know that, but I don't. Let me get the manager." She appeared, heard our question, and went back to her office for the words she obviously didn't know. On her return she said, "We're not supposed to give out that information." We mentioned that it's been published many times; could she perhaps give us the number of someone who might know it? She directed us to the local franchiser. He promptly gave us the answer, after his secretary first told us she didn't know it either. What good is a philosophy or purpose statement if nobody knows it?

Management everywhere fails to communicate the core purpose of their business in clear, effective, multimedia fashion. Communication failures with employees ultimately fail the customers. And there goes the whole ball game. With their purpose undefined, not communicated, or misunderstood, employees often deliver inconsistent or unpredictable results.

We are about to explore the inner workings of an aircraft maintenance case in written communication—an example of the life-or-death role that written communication can play. If you want to skip ahead to the description of the accident at Dryden, Ontario, go ahead. All you really need to know for now, for the whole chapter in fact, is that good written communication is absolutely necessary *and* that it must be accompanied with other good communication, spoken and unspoken.

To follow our logic, let's look at McDonald's again.

Like most people, we've never formally studied McDonald's. But like everyone, we know the company from the outside, so we can draw some useful conclusions just from what we already know.

First, it's not happenstance that McDonald's makes heavy use of marketing, public relations, advertising, training, and internal communication *and* that they stand at the pinnacle of the fast-food business. This is not a coincidence. It's cause and effect.

The most important part of McDonald's communication success, of course, lies in the product itself. Nothing communicates so perfectly as what you actually do, regardless of what you say about it. McDonald's meets the expectations of their customers.

All other communication, inside and outside the drive-up window, must focus on delivering the goods, as promised, starting with the food but also including

Mechanics Cannot Thrive on the Written Word Alone

the QSCV they've promised. That takes a lot of doing on everyone's part, and only good communication across the board can keep it all on track.

Good communication almost always means multimedia communication—that is, using several channels to get across the same message. The core message to employees lies in the company's actual aims versus their stated aims. If management's fine words and actions match, and if the employees are committed to the same purpose, they meet the test for excellence in communication: shared purpose.

Managers everywhere say they want good communication. But that's the talk. Does management walk their talk? Do they use multimedia to get their message across? Do they make sure that employee feedback actually gets back to them? Any gap between words and deeds confuses, frustrates, and de-energizes the workforce.

Missed communication opportunities cause big problems. But good communication pays back with very good results. At McDonald's, for example, the invention of a whole new fast-food meal, featuring the Egg McMuffin, reportedly came about because McDonald's solicited ideas on that very topic and one individual *outside* the company brought in his breakfast sandwich for show-and-tell.

Being open to new messages, even when they come from the outside—that's good communication, leads to shared purpose.

When communication about maintenance safety, productivity, and cost control fail to fulfill the company's central purpose, communication may disintegrate into chaos.

A bad meeting, a bad newsletter, a bad new hire, a bad policy, a bad training program, or a bad workplace environment—these are examples of communication gone bad. In effect, they demonstrate that the company's stated purpose is not its real purpose. So what is? Can you imagine what these hapless employees might think management's true purpose is? The possibilities are endless, from the merely off-target to the seriously counterproductive.

Which of these conflicting purposes will the employees follow? Luckily, most maintenance mechanics are strong enough, even under pressure to do otherwise, that they will follow their first purpose: safety. But many others feel they are not always supported in pursuing this core purpose.

Without clarity of direction and suspicious of management's aims, employees at every level won't know for sure just what is important. They'll tend to attack each other instead of joining forces for their shared purpose, the most important part of which is satisfying their customers' safety needs. Keeping them alive. Written communication must be backed up with every other kind of word and deed, to make the safety message crystal clear. Time and cost are important, but safety is number one.

When management's words and deeds don't match this message, it can run up the cost of everything. It certainly runs up the risk of errors. When managers' words and deeds don't match, employees may conclude that they're indifferent, or inconsiderate, disloyal, incompetent, manipulative, dishonest, or worse. Managers' worst motives are suspected. Whether mangers truly are indifferent or merely poorly skilled communicators, the painful consequences are the same.

To communicate well and to *do* well, the core purpose of the airline and its employees and customers must be clear as mountain spring water, unmuddied with conflicting messages, whether spoken or unspoken, written or oral. Written communication must always reflect the company's core purpose. And it must serve as the official record for all the other forms of work-related communication.

Other forms of communication have their place; but one thing is paramount: They must be aligned with the written statements, the unwritten statements, and the "statements" seen in management's behavior. Incongruence brings chaos.

All stakeholders must hear this same core purpose repeatedly, through many media. Not just employees, but employees' families, community neighbors, suppliers, government regulators—literally everybody who has a say in whether the enterprise stands or falls. Performance rises in direct proportion to the quality of the company's communication.

Consider the "media" used by McDonald's to get their message across: store appearance and location, employee training, and their on-the-job appearance and behavior. McDonald's also uses store signage, billboards, TV ads, toys, and jingles. They use uniforms, their own Hamburger U., Ronald McDonald, point-of-purchase displays, posters, games, movie tie-ins, and the Ronald McDonald House for the parents of children hospitalized nearby. They do all of this and more. They don't just sell hamburgers. They sell QSCV, and they sell it in various ways.

McDonald's purpose knits these components together. The company makes sure that its purpose comes across clearly, honestly, and well. It presents a seamless and consistent message to every stakeholder, in every role they play. Multimedia reinforce and advance McDonald's purpose. But an important link, when missing, can defeat the effects of great posters, training programs, and all the rest. Management must always reinforce their purpose in words as well as in their behavior. The fact that some employees in our opening example didn't know what QSCV is means that even for a hugely successful company like McDonald's, communication can *always* be improved.

The media for gaining the kind of commitment we are talking about are varied and mutually supportive. The core message is always the same, but the messengers must change to suit the occasion. Several media must be used to drive the message home. A single bulletin or handbook or letter cannot ever do the communication job on its own.

People require a multimedia approach partly because that's what we've come to expect, thanks to big marketing, PR, and advertising efforts all around us. And we are ourselves multimedia creatures. Our nature is to look at a thing from all directions in order to build a complete, accurate, and trustworthy guide for decision making and action.

Oral Communication Has Limitations

When we think of communication, we ordinarily think of people talking to one another. Aircraft maintenance people don't talk productively with each other enough.

AMTs usually get their job assignments face-to-face from their supervisors or from lead mechanics. Oral communication is also the primary mode of communication in on-the-job-training (OJT) and in classroom discussions. Likewise, at breaktimes, lunch periods, and at quiet times in the maintenance ready room, mechanics talk with each other. And talk is important.

But let's make it effective talk.

The most important improvement in oral communication to date has come with the teaching of assertive communication styles (assertive, not aggressive or passive). These new skills enable AMTs and their managers to effectively confront one another about unsafe acts, misunderstandings, and errors. It also enables them to hear others effectively when they themselves are assertively confronted.

But important as it is, plain talk is not the only important model for human communication. Research evidence has shown that nonverbal cues—body language and facial expression—can be powerful vehicles for communication [Ref. 1].

Here's to you, maintenance: It takes more than a bulletin to get the message delivered. Don't stop writing bulletins, but go multimedia for best results.

Written Communication Is As Important As Ever

But the spoken word and body language together do not cover all of the most important communication in aviation maintenance. Despite the central importance of face-to-face communication, written communication remains as important as ever. Maybe more than ever. So getting clear on the *purpose* of any endeavor is the first, most important ingredient for success. Writing it down helps make it clear, and clarity avoids misunderstandings and mistakes, some of which can be very, very costly.

Written communication plays a central role in every communication system. Written communication among mechanics, with their management, and with others outside maintenance is of vital importance. And written communication is more than just record-keeping. Written communication is the bedrock of all communication in maintenance, supporting the various messages delivered otherwise. But it has to be done well, just as well as the maintenance itself.

Written communication tells mechanics how to do the job. If they read it. If they understand it. And if they don't have any questions. Face-to-face communication is ordinarily used to solve those three problems. Thus, the intelligent response to these well-known problems is to support written messages with face-to-face communication. We have to select the right words.

The following case is a study in poor communication—written, spoken, and unspoken—where one communication error after another set off a long chain of events that culminated in the death of 24 people. After the crash, the Canadian Commission of Inquiry reviewed the chain of events and made some 200 specific recommendations for the company, the industry, and the regulators. This one crash uncovered numerous links to disaster across the entire air transport safety system.

The Accident at Dryden, Ontario

In the snowy midday of March 10, 1989, a regional airliner attempted to take off from the airport at Dryden, Ontario. It didn't make it. There were 69 people on board.

Twenty-four of them died.

The official hearing found that the captain took off with ice on the wings of his plane. The story might stop there, except that the Canadian Commission of Inquiry looked more closely and uncovered many weaknesses in the air transport safety system.

Communication problems led the list.

The airplane was a Fokker F-28. It was flown on Flight 1363 by Air Ontario. The young company was a merger of two smaller companies that had operated in very different circumstances, one in the more populated areas of southern Canada and the other in the bush-flying conditions of Canada's northern regions. The F-28 was part of a new fleet the company had just acquired. The Fokkers were the first turbojets the company had ever owned. To operate effectively, Air Ontario had to manage its unfamiliarity with the new planes. Unfamiliarity always risks poor practices and poor results.

To this mix, yet another layer of complexity was added when Air Canada, a much larger carrier, acquired a controlling interest in the company. Thus, for employees at Air Ontario in early 1989, many operating procedures and policies were new to them, not to mention the new planes and new colleagues. Add these components to the normal complexity of commercial air transport, and the risk of mistakes increases exponentially. The Dryden accident is a case in point.

The captain and crew were on the last leg of their journey. The flight began in Winnipeg, Manitoba, and completed scheduled stops in Dryden and Thunder Bay, Ontario, and again at Dryden on the return trip to Winnipeg. It was a holiday weekend and the flight was full—some passengers had been diverted from a canceled flight, and the crew knew the passengers were tired and frustrated.

In the immediate events leading up to the crash, Flight 1363 had been delayed at Thunder Bay to take on the additional passengers, but had incurred another delay for off-loading fuel to adjust the new weight and to balance the aircraft. The flight was over an hour late for the second scheduled arrival at Dryden that day.

At Dryden, during this second stop there, Flight 1363 required yet another delay to take on more fuel. It was snowing lightly. Refueling with the engines running is a questionable safety practice, but since it was not prohibited by policy or law at the time, the captain kept the engines running and used them for electrical power and cabin heating during this stop. But there was a more important reason the F-28's engines remained running. The little Dryden airport had no air-start facilities, and on a red paper placard covering the pilot's airstart control switch, the note "INOP" meant that the plane's auxiliary power unit (APU) was inoperative. The captain knew from that information that if he turned his engines off he wouldn't be able to start them again that afternoon.

Before the roll-out for takeoff from Dryden, the captain queried Dryden ground personnel if they had deicing capability and whether it was needed. He was told that deicing was available but that it didn't appear to be necessary. The captain was an experienced bush pilot who had flown into and out of many fierce winter storms in his career. Also, he knew the company rule that airport deicing was only permitted with the engines off; "hot deicing" was not allowed.

The company had no policy about operating without an APU. The Canadian Commission of Inquiry pointed to this glaring absence: "It would be expected [as part of the larger organization formed by the Air Canada acquisition of Air Ontario] that there would be a great deal of effort made to create universal policy and procedure. However, as parent company to Air Ontario, Air Canada did not involve itself in the creation of the proposed F-28 MEL ["MEL" refers to a company's repair deferral process]. Had Air Canada involved itself in the F-28 MEL process, it surely would have included its own guideline restricting aircraft from landing at facilities lacking ground start-up units when the aircraft's APU is inoperable." [Ref. 2 V2, p. 504–506]. As it was, there was not a policy, written or otherwise, to forbid such scheduled stops.

After waiting with the engines on and with snow building up, the captain proceeded to taxi toward the runway, but was held from taking off by the control tower while a small private plane approached to land—a Cessna in trouble in the snow squall that had just come up. That squall dropped a good deal of heavy snow. During the seven minutes the F-28 waited for the smaller plane, a buildup of snow on its wings was noticed by several passengers, but was apparently not reported to the cockpit. The Cessna landed safely, and at 12:09 P.M. Flight 1363 continued to taxi to the runway and accelerated for takeoff. The F-28 lifted only to tree-top level before it crashed some 900 meters from the end of the runway. Deadly icing was the immediate cause.

You may ask, what was the role of maintenance in this accident and what was the importance of maintenance communication practices?

A number of recurring problems were associated with the aircraft's APU. On six different occasions—January 21, February 27, March 5, March 6, and twice on March 8—an oily smoke, smell, or haze was reported in the cabin. Furthermore, both a captain and a maintenance supervisor noted, at different times, deficiencies in air pressure and an "oily smell."

On the morning of March 9 (the day before the accident), a maintenance crew chief at the company's Toronto maintenance base reviewed the aircraft's journey log to discover notations by the aircraft's two previous pilots regarding issues pertaining to the aircraft's APU. The first pilot noted that "APU air pressure is low." The second pilot noted that "engine starts are becoming more and more difficult." The crew chief and another mechanic performed some trouble-shooting. They tried a new load-control valve, but when that didn't improve performance, the original load control valve was reinstalled and the APU was started successfully. At this time, both the aircraft's APU and fire detection system tested as functional. After the APU was tested, its fire shield was reinstalled by a trainee.

At approximately 4:00 P.M. that afternoon, another crew chief beginning start-up procedures on the aircraft found the APU fire-detection system was not operational. Searching for the cause of the trouble, the crew chief found a loose wire, but reattaching it did not solve the problem. During the accident investigation, it was speculated that when the trainee had earlier tightened the fire shield for the last time, he might have pinched a wire in the fire-detection loop. If this error occurred, it could have rendered the fire-detection system unserviceable.

Air Ontario maintenance was faced with two options at this time: Ground the aircraft until the problem was rectified or defer maintenance of the APU fire-detection system. It was decided to defer maintenance, following the section of the MEL dealing with the APU fire-extinguishing system. This decision was discussed by the crew chief, a lead inspector, and the pilot who was scheduled to fly the aircraft from Toronto. The matter was also discussed with both the company's systems control and maintenance control personnel in London, Ontario. The maintenance crew chief noted in the aircraft journey log, "APU will not fire-test." He added in the "defect modified" section, "Deferred as per MEL 49-04." After making the journal entry, he then put up the red placard on the cockpit APU panel that read "INOP."

On the afternoon of March 9, the pilot accepted the aircraft with deferred maintenance on the APU fire-extinguishing system. In addition, he used the APU to start the aircraft engines in Toronto prior to his departure. This was permitted by MEL section 49-04, which required that, with an inoperative fire-extinguishing system during start-up, the captain must have any passengers on board disembark. He was then to arrange for constant fire monitoring by the ground crew. The plane was flown to Winnipeg that Thursday evening and was waiting there for the Friday morning flight to Dryden, to Thunder Bay, and back to Dryden where the accident occurred.

Why Did Maintenance Personnel Make the Decisions They Did?

Because they lacked experience with the F-28 aircraft and its systems. F-28s had been a part of Air Ontario's fleet for only eight months. Although the personnel involved had not worked on F-28s before, they had attended two weeks of classroom training at another airline. Eight hours of that training was on the operation of the APU, of which three to four hours were devoted to the fire-detection/extinguishing system.

On March 9th maintenance was not aware of where the aircraft would be operating next and of the possible consequences of an inoperative APU. The crew chief and lead inspector discussed the APU problem with the pilot taking the aircraft as well as with systems and maintenance control operations in London, Ontario, but they, too, apparently lacked appropriate awareness of the situation ahead.

The importance of written communication is illustrated in this case. There was no MEL for "fire detection"—the only choices were MELs for just the APU (49-01), *or* for just "fire extinguishing" (49-04). Neither option covered the contingencies in this case. Two qualified mechanics and a pilot had chosen the latter MEL. The MEL placard the mechanics placed on the APU controls was a study in poor written communication—it was vague, limited, and misleading.

Perhaps they should have MEL'ed the APU (49-01) and really made it legally impossible to operate the APU. But they also should have known that an aircraft without APU could not be started at Dryden and thus that the aircraft should be grounded instead of having to make a scheduled stop at that station.

Instead they did the reverse, they allowed the airplane to leave without warning future pilots that the APU *could be used* to start the engines under the very special circumstances of the MEL they chose (49-04).

The placard on the APU controls could have said, "Would not fire-test—APU can be used to start engines under conditions of MEL 49-04." That would have been more useful to the pilots!

Encouraging Attempts to Improve Written Communication

A "Maintenance-Record Speedometer System" Works

One very effective program to improve descriptions on work records was implemented in a large repair station [Ref. 3]. The program, which helped reduce the number of imprecise abbreviations and phrases, began with an investigation of the repair station's work records. The investigators found that for a twelve-month period 40% of them contained vague, ambiguous, or abbreviated phrases not in keeping with the intent of the FARs. Employees surveyed during the investigation reported that the repair station's records had been an ongoing problem and that supervisory enforcement of policy was cyclical and short-lived. The resulting management "crackdowns" had not been very effective. Error rates that went down during the crackdowns returned to previous levels when scrutiny was removed.

Investigators recommended that work-record error rates be posted daily on the hangar floor. Simple posters were created and posted in a central location next to the FAR requirement posters. That day's performance, compared with historical data, was immediately available. Mechanics could see within hours the failure rate of their input. The results were remarkable. Within eight weeks the 40% error rate went to zero, and it did not return to previous levels.

The author of the study remarks that in his experience "organizations that continuously measure their performance against their goals always seem to achieve their goals and usually much more" [Ref. 3]. Driving the measurement down to the shop floor seemed to make FAR-compliant maintenance records occur with very little management involvement—they didn't need continuous management monitoring once the new measurement and feedback system was in place.

Employee Participation Improves Paperwork

Chapter 2 reported the causes for paperwork errors as listed by one company's foremen, leads, and mechanics. Ten separate causes formed a common set among some 160 people interviewed [Ref. 4]. These causes included trouble understanding written instructions and documents, antiquated technology, inadequate training and preparation, pressures and distractions from management and other departments, plus poor communication with upper management and with other maintenance sections.

Since 1992, the company, the mechanics' union, and local FAA regulators have monitored paperwork errors and supported one another in a series of improvement efforts. They are alert to discovering new avenues to improving paperwork processes and reducing paperwork errors [Ref. 5]. This extensive intervention is more completely described in Chapter 9.

The Ergonomics of Good Forms and Documents

"Ergonomics" refers to the design of tools, equipment, and machinery in which the human body, its shape, and its limitations guides the design of equipment humans use. Much research into human physiology and anatomy supports this approach. Mechanics' work cards, aircraft log books, and manufacturers' repair manuals are a kind of technology too. Recent research has shown that ergonomic design of such forms and documents reduces error rates, thus improving both safety and productivity.

Professor Colin Drury and his associates at New York's University of Buffalo have been investigating the ergonomics of forms design in aviation maintenance. Their research has resulted in significant and relevant findings. The Buffalo group has shown that improving the design and layout of work cards and log books is as effective as computerizing those same forms. For instance, they have illustrated that work cards printed in all capital letters (a common practice in "CITEXT" computer printouts of work cards) are more difficult to read and are easier to misinterpret than work cards using upper- and lowercase letters [Ref. 6]. They have also established that the sequence of tasks on a work card can be optimally arranged to increase efficiency: A poorly designed aircraft overnight check form might have the mechanic first checking tire pressure, then cabin lights, then the brakes, and so on. Reordering tasks in better sequences has been shown to reduce errors and improve productivity [Ref. 7]. The Buffalo group has also demonstrated the effectiveness of using "Simplified English" (a proposed standard language for aircraft maintenance throughout the world) in reducing errors and misunderstandings even among native English speakers in the United States [Ref. 8].

Good Personal Communication, plus Write-Ups, Improves Performance

Sometimes what is written cannot be improved upon quickly enough, and other steps must be taken to improve the information transfer. When mechanics use written information prepared by other departments, they may need to intensify their communication efforts. A laboratory study done by faculty and students from Purdue University illustrates this point. These researchers found that direct, face-to-face communication of log book write-ups ("squawks") improves mechanics' troubleshooting and reduces repair time [Ref. 9]. Two scenarios were tested in this study.

Student pilots in a simulator for a Boeing 727 were given a maintenance discrepancy and were directed to write it up in the log book as if they were on an actual flight. In one scenario, the flight crew simply handed their log book write-up to the student maintenance technician, who was directed to identify and repair the corresponding discrepancy on an actual 727-100 aircraft. In the other scenario, the maintenance personnel went to the flight deck and discussed the log book write-up with the flight crew prior to being directed to identify and repair the discrepancy.

The investigators found dramatic differences between the two conditions. The student technicians relying solely on the written log book write-up spent more time troubleshooting, and half were unable to identify (and therefore repair) the flaw. For those student technicians who had discussed the squawk with the simulator flight crew, all were able to go directly to the problem and make the necessary repairs in a short period of time. Comments from the experimental subjects included such statements as, "We need to do more of this, more information is communicated when we talk about it," and "I gained a much better understanding of the pilot's/mechanic's job" [Ref. 9].

A Parting Thought

Written communication in aviation maintenance is at the very heart of the system. Not only is it required by U.S. law (FARs); good information documentation is the backbone of good safety practice. Written communication facilitates people communicating with one another over long distances and over time. But it is only one tool in the communication tool chest. It should not be relied on alone. Ideally, the writer is known to the reader, and the two discuss the subject matter directly. Because that is not always possible, the writer needs to have a broad understanding of the situation as well as an appreciation for the reader's point of view; written information must be understandable, unambiguous, and clearly relevant.

References

[1] Mehrabian, A. *Silent Messages: Implicit Communication of Emotions and Attitudes,* 2nd ed. Belmont, CA: Belmont Press, 1981.

[2] Moshansky, V.P. *Commission of Inquiry into the Air Ontario Accident at Dryden, Ontario* (Final Report, Vol. 1–4). Ottawa, ON: Minister of Supply and Services, Canada, 1992.

[3] Hutchinson, III, C.R. "Aviation speedometers, metrics on the hangar floor." *Ground Effects* (Jan.–Feb. 1997): 1–5.

[4] Taylor, J.C. *Maintenance Resource Management (MRM) in Commercial Aviation: Reducing Errors in Aircraft Maintenance Documentation* (Final Report of 1993–1994 Project Work). Sociotechnical Design Consultants, Inc. Box 163, Pacific Palisades, CA 90272, December 1994.

[5] Kania, J.R. "Panel presentation on airline maintenance human factors." *Proceedings of the 10th Meeting on Human Factors Issues in Aircraft Maintenance and Inspection.* Washington D.C.: FAA, Office of Aviation Medicine (1996).

[6] Patel, S., P.V. Prabhu, and C.G. Drury. "Design of work cards." *Human Factors in Aviation Maintenance—Phase Three, Volume 1 Progress Report,* DOT/FAA/AM-93/15, Springfield, VA: National Technical Information Service, 1993.

[7] Patel, S., C.G. Drury, and J. Lofgren. "Design of work cards for aircraft inspection." *Applied Ergonomics* 25(5), 1994: 283–293.

[8] Chervak, S., C.G. Drury, and J.P. Oulette. Field evaluation of Simplified English for aircraft work cards. *Human Factors in Aviation Maintenance—Phase Six, Progress Report.* Springfield, VA: National Technical Information Service, 1996.

[9] Wulle, B. and M. Lapacek. "Effective pilot/technician communication aids troubleshooting." *Ground Effects* (May/June 1997): 10–11.

Chapter 9

MRM Is Multiparty Cooperation, Open Communication, and Error Reduction

OVERVIEW: MRM—A Strong Model for Multiparty Communication, Cooperation, and Error Reduction

In 1992, USAir (now known as "USAirways") introduced a bottom-up approach to maintenance resource management (MRM). A trade union together with the company's maintenance management and local FAA regulators created MRM to better understand a serious problem they shared and to seek ways to cure it at its source. The problem was a high incidence of maintenance-related paperwork discrepancies. A series of mergers with other airlines had increased the size and complexity of USAirways paperwork problems. Among other things, this increased complexity led to confusion in signoff procedures and compliance with policy. During the early 1990s, the number of paperwork discrepancies rose dramatically.

The International Association of Machinists and Aerospace Workers (IAM/AW) wished to protect and serve its members by eliminating sources of paperwork errors for which the members might receive disciplinary action. Their mutual goal was to ensure better compliance with policy. Union representatives approached USAirways management in Pittsburgh (the company's maintenance headquarters) and the local FAA Flight Standards District Office (FSDO 19) on the subject of paperwork error reduction.

A "win-win-win" attitude among the three parties was apparent from the start. Everyone had something to gain—the FAA and the IAM both wanted fewer enforcement actions, and the FAA and USAirways wanted improved documentation performance. The principals reasoned that if errors could be reduced or eliminated through better communication and collaboration, then punishment might seldom be necessary. The three parties took inspiration from the successful 1990 "Altitude Awareness" test program, a narrowly focused study to reduce flight-altitude deviations. That program had achieved good results through the collaboration of the Air Line Pilots' Association (ALPA), USAirways management, and the Pittsburgh Flight Standards District Office (FSDO). So an example of successful three-party collaboration was right before them, in their own industry.

The Bottom-Up Research Program

The error-reduction "group of three" reasoned that AMTs wanted to work safely and would collaborate with others to do so, but they lacked the administrative skills, or the trust, or the communication skills, or the opportunity for successful collaboration. A research program was launched to test the effects of the three parties directly approaching AMTs and their foremen for help in reducing problems with paperwork. This was presented as an important issue, even though paperwork compliance had not usually been understood as safety-related, as an important element in maintenance error control.

The research program was the creation of the three parties (the IAM/AW District 141 flight safety coordinator, the USAirways quality assurance manager, and the FAA local FSDO19 principal maintenance inspector) plus their facilitator, an organizational psychologist. In the remainder of this chapter, we will refer to this extended group as the "MRM Research Team."

The project followed the logic of "action research" [Ref. 1; 2], a high-participation form of organization intervention in which there is no action without research and no research without action. By definition, action research also means that both the research and the action are carried out largely by the people involved. This is a powerful means to creating organizational change.

The action research project used as its metaphor a funnel, turned upside-down. Usually, we throw our problems into the wide top of the problem-solving funnel, and we get a narrow solution out the bottom. In this case, however, paperwork problems were put in the narrow part of the funnel, and a wide range of general maintenance solutions came out of the wide part. The team knew from the start that the issues of documentation would inevitably touch on other problems too. Record-keeping is not just paperwork. It's part and parcel of the work itself.

The MRM Research Team believed that rising or even static error rates or discrepancies in maintenance paperwork in some cases are a sign of lowered motivation and/or low trust within the maintenance system. This, in turn, leads to lowered quality of the maintenance work overall. The objectives of MRM were therefore defined as follows:

1. To reduce maintenance-related paperwork errors and thereby to improve maintenance quality
2. To build the trust for better communication and collaboration among members of the maintenance system and with the FAA

The MRM planners decided early on to begin with the people who produced the errors. They therefore involved mechanics, inspectors, leads, foremen, and higher management in the search for causes, intervening circumstances, and possible solutions.

The Company's Independent Efforts to Improve Aircraft Documentation Quality

USAirway's ongoing efforts to improve its record of paperwork errors predate MRM. As early as 1989, its managers were concerned about the high incidence of paperwork errors following its mergers. Three subsequent changes resulted.

First, the aircraft records department *improved its audit productivity.* In 1989 and 1990, audit quality rose by increasing the number of records clerks. During 1991 and early 1992 (preceding MRM), these new records clerks increased their audit skills and their productivity in finding errors and reporting them to maintenance foremen and mechanics. After 1992, the aircraft records department productivity continued to improve, despite a reduction in clerks and a tightening of its audit and quality standards.

Second, the company *changed its signatory requirements* to include employee number and to insist on legible signatures. Unfortunately, this positive change in procedure opened a new avenue for errors. The change made it easier to identify those responsible for unclear signatures, but the new employee ID numbers could come up illegible as well.

In response, the company made a third change in 1994, which was to *require use of stamps,* instead of signatures, on all forms except the aircraft log book. Although these changes affected error rates in various ways, none of them addressed the issues of trust, open communication, and partnership. We'll get to that with the introduction of Maintenance Resource Management (MRM).

PHASE ONE: Participative Data Collection

The First Year of MRM

MRM's first-year efforts in 1992–1993 used improved problem-solving communication techniques with USAirways' AMTs, leads, and foremen [Ref. 3; 4]. The company made a special effort to measure and document the accuracy of maintenance paperwork completed by AMTs and their foremen. These performance measures were used to evaluate the success of the program, and it was a success.

During 1992, some 100 AMTs and foremen in four large line stations were brought together in 18 focus groups, convened by the MRM Research Team and guided by its outside facilitator. The focus groups were organized with AMTs in one group and foremen and lead mechanics together in the other groups. In early 1993, twelve additional focus groups were convened in three smaller line stations and in staff support units.

The focus group discussions were conducted on all shifts—afternoon, night, and day shifts at the four large line stations. The MRM Research Team usually began their visit at about 10 p.m., during the conclusion of the line maintenance second shift. The visit began with the station manager or designate gathering all available AMTs, leads, and foremen. Also present were the IAM flight safety representative, the USAirways quality assurance manager, and the FAA principal maintenance inspector. These three would introduce themselves, the outside facilitator, and the motives behind the MRM program. They then asked for questions and invited volunteers to join a focused discussion group of AMTs, or a similar group of leads and foremen. In a typical station visit, MRM focus groups of five or six AMTs met on each shift, and one equal-sized session of leads and foremen was convened as well. They were guided and led by the same facilitator, and the sessions lasted about two hours each.

During their discussions, participants were asked directly and in confidence to list causes of paperwork errors and what they thought would reduce them. These people then worked through the two lists. This process frequently revealed substantial agreement among participants. Most of their proposed solutions were later implemented, and paperwork errors were subsequently reduced to various degrees, depending on the specific changes made.

Initial Results of MRM Focus Groups

The MRM Research Team reviewed the results from the 18 initial focus groups. They discovered substantial agreement across the lists of error sources and suggested solutions [Ref. 3]. The MRM Research Team then created the summary lists shown in Table 2.5 in Chapter 2 and Table 9.1. The final list of most important paperwork error sources was confirmed by the dozen subsequent focus groups in early 1993.

The list contains a number of communication-related issues: poor AMT-management communication and maintenance system technical information, little use of "best communication practices" from the companies acquired in the mergers, overwhelming information requirements on the flight line, poor document design, less than adequate engineering information, inadequate training in paperwork, and antiquated information technology. Table 9.1 lists the final recommendations.

In February 1993, after the first nine months of the project, a preliminary report was prepared and distributed among the company's maintenance managers, IAM union officials, and the local FAA FSDO managers. Reviewing the report, senior management questioned the applicability of the recommendations to smaller stations, and they wondered whether engineers' and records clerks' views were similar to those of the mechanics.

TABLE 9.1 MRM FOCUS GROUP HIGHLIGHTS

Error Causes Specified	Solutions Suggested
◆ Poor AMT-management communication about technical information	Increase two-way AMT-management communication about technical matters
◆ Antiquated information technology (e.g., computers, faxes, microfilm readers)	Install newer film readers and a real-time computerized log book
◆ Poor document design or clarity and confusing or conflicting technical information	Involve AMTs in revising the paper log book form and in rewriting the General Maintenance Manual
◆ Complex/redundant engineering information	Provide AMTs' review and feedback to engineers before Engineering Orders are issued
◆ Overwhelming information requirements for through-flights	Reduce the number of maintenance checks (and their paperwork) on through-flights during the day
◆ Inadequate training in paperwork use and practice	Provide AMT training in paperwork—including OJT and on-shift meetings by lead or foremen

Source: J.C. Taylor, [Ref. 3]

Between February and May 1993, another 60 maintenance respondents in three smaller line stations participated in ten focus group interviews about paperwork problems and possible solutions. Engineers and aircraft records clerks at the USAirways maintenance headquarters participated in two additional focus groups. The findings from the new discussions again confirmed a consensus on the causes and solutions for paperwork errors.

The Second Year of MRM

During the latter part of the second year, the company (with the support of the IAM and FSDO) made several changes suggested by the MRM focus group participants. Except by guesswork, most of the mechanics affected did not know that the changes were the direct result of MRM focus group recommendations. This lack of linking information resulted not from intent but from simple oversight and omission. Two years would pass before the company would attempt to widely publicize the MRM approach, its programs, and its benefits [Ref. 5].

PHASE TWO: Making Changes With and Without AMT Involvement

Formal Training Conducted

The focus groups had recommended that AMTs get formal training in the use of required forms and job cards. This proposal was acted on first. The company, with concurrence of the IAM and the FAA FSDO, designed and implemented a paperwork training course for all AMTs in line maintenance. The paperwork training was conducted by USAirways technical training department personnel and was delivered between July and September 1993 to 1,300 AMTs in 37 line stations.

The research report published 18 months later documented the degree to which paperwork errors diminished immediately and then remained lower for nearly a year, before approaching previous levels [Ref. 3]. That same report showed that in a sample of line maintenance mechanics interviewed six to eight months after the training, few of them remembered the training, but new hires in the sample wished that they had received such training even earlier, during indoctrination. These combined findings show that improving the purely technical content of training can and does improve safety, but the improvement is also short-lived and requires additional effort to sustain lasting change.

Opening Communication with Upper Management

At the same time as the paperwork training, the USAirways quality assurance department installed a telephone hotline as another effort to improve communication between AMTs and management. This change addressed the first recommendation in Table 9.1. The hotline was intended for all air worthiness flight-safety issues of a nonemergency and nongrievance nature, and anyone in the company could use it.

There was no formal plan for the company to respond publicly to hotline issues. Despite that, the opportunity to speak directly to management produced positive results for safety communication. The hotline resulted in about ten calls per month during the last six months of 1993. The 1994 hotline showed a lower rate of about four calls per month, with line mechanics the primary callers. As calls to the company's hotline continued, calls to the FAA's hotline decreased dramatically [Ref. 4], and the FAA's calls from USAir mechanics have remained low ever since. This indicates that more problems are being resolved from within, among the people directly involved.

Wider Communication with Preshift Crew Briefings

Following the second round of focus group interviews in early 1993, USAirways management and the IAM union agreed to implement preshift meetings in one small line station and to have volunteers from that station redesign the company's aircraft log book. In August 1993, the shift foremen at that station received training in conducting briefings, and they held short meetings at the start of each shift for a number of months thereafter.

This initial effort to encourage team meetings produced promising results. Ten months of paperwork performance of the station where meetings were held were compared with that of an equivalent station that did not hold meetings. The meetings did indeed lead to a noticeably lower error rate for the experimental station; this was not observed either in the control station or in any other USAirways station for the same period [Ref. 3 p. 20].

During 1992 and early 1993 (before the shift meetings began), the morale and service reputation of the experimental station were below par, while the control station appeared to be better. Following the introduction of shift meetings and involvement in revising the log book, maintenance management reported that pilots' complaints about the experimental station dropped and remained low. Indeed, the chief pilot told the senior maintenance vice president that "Whatever you're doing at that line station has made their service much better. Keep it up."

In subsequent evaluations of the shift briefings, however, mechanics in the experimental station complained that two-way communication still was not encouraged, the meetings were not supported by the station manager, and mechanics and foremen were not receiving timely feedback about actions taken as a result of the meetings.

The foremen also reported weaknesses in the shift briefing program. They needed help creating agendas, managing questions in a group setting, and finding a source for company information during afternoon and night shifts. The foremen thought they would benefit from helping to design and implement appropriate continuation training in the leadership of meetings [Ref. 3], an example of action research at work.

Unfortunately, this encouraging beginning did not last. With foreman turnover and transfers from the station, and with little support from the station manager, the preshift briefings gradually diminished in frequency, and by August 1994 they were only a memory. Management had failed to respond to the need to maintain and improve the program.

But it wasn't the end. The idea of team meetings had really only just begun. Beginning in 1995, USAirways implemented a systemwide program of "self-directed teams," intended to foster multiparty (rather than one-way) communication. However, these team meetings have not, during their first two years of operation, discussed safety-related issues, an oversight that should be addressed.

During 1993 and 1994, the experimental line station did succeed with its own preshift meetings, however briefly, but mechanics there also completely revamped the company's aircraft log book.

AMTs Design Their Own Paperwork

The MRM focus groups of 1992 and 1993 complained about poor document design, and they recommended that AMTs be involved in redesigning the forms. Beginning in August 1993, USAirways, for the first time ever, asked their AMTs to help improve the forms and documents they work with daily.

Mechanics Create New Log Book

In 1993, the company had two log books. They had a wide-body log book for ETOPS aircraft (two-engine planes designated for transoceanic service), and they had a log book for all other airplanes in their fleet. These two log books both worked "OK," but the MRM focus groups had pointed out the unnecessary differences between the two log books and the added complexity of working with two books instead of one. USAirways was by that time a couple of years beyond the last of several mergers, and consolidating the legacy forms seemed appropriate.

The quality assurance (QA) manager informally approached the New York LaGuardia line station AMTs about redesigning the log book, but he was met with skepticism. The LaGuardia AMTs told the QA manager, in essence: "Get outta here. This is just a one-shot deal. You guys in Pittsburgh are not really interested in fixing this log book."

The manager persisted, however, and he kept visiting the station—on night and afternoon shifts as well as during the day—sitting down in the break area with the two log book forms and asking AMTs and lead mechanics how they would improve and consolidate them. This manager became well-known at LaGuardia during the fall of 1993. It was not long before AMTs there were calling him in Pittsburgh, asking him to drop by so they could talk further. In the end they completely revamped the log book [Ref. 4]. Of course, they consulted with maintenance management and the technical departments, but it was their log book.

In previous log books going back 33 years, something was always wrong from the mechanics' point of view. When the new, employee-designed log book was issued in April 1994, it was enthusiastically received throughout the system and has remained the accepted standard to the present. When further opportunities for improvements are presented, US Airways now knows how to get the job done right. They now have a new "institutional memory" of effective participation.

The Third, Fourth, and Fifth Years of MRM

USAirways AMTs Rewrite General Maintenance Manual

In the focus groups of 1992–1993, both AMTs and foremen repeatedly stated that the General Maintenance Manual (GMM) wasn't easy to use or understand, and the manual itself thus contributed to paperwork errors. The GMM (and its companion administrative manual) contained basic policies that affect mechanics and aircraft maintenance in a general nature not necessarily confined to specific types of airplanes.

For 40 years, USAirways and its predecessor companies had policy manuals where original policies were held, into which new ones were added and cross-referenced, and from which nothing ever seemed to be removed. In short, a mess. Despite the complex and convoluted nature of these evolving manuals, mechanics and foremen were simply expected to comprehend and comply with them.

USAirways had such good success with the log book redesign that management and the union decided to go a step further, and have AMTs throughout the company rewrite the policy manuals. In July 1994, the company and the union announced plans to restructure the policy manuals. All members of maintenance operations were invited to provide input about problems with the manuals and suggestions for improvements. A steering committee was formed from members of the MRM Research Team, the IAM, the QA department, and the technical publications, maintenance programs and reliability departments. That committee organized each page of the manual into sections such as "deferred maintenance," and "time cards," and then sent the sections out to volunteers in maintenance stations throughout the system. The instructions were simple: "Here folks, you wanted to do it. Reformat this section so that it is user-friendly and send it back to us." The rewritten sections were combined into one manual called Maintenance Policies and Procedures (MPP). The MPP was released for use in May 1996. It had been totally revised by its intended users and other interested parties. Nearly 80% of the people involved in rewriting the MPP were IAM members [Ref. 5].

Reintroducing Computerized Log Books

The MRM focus groups of 1992–1993 identified problems AMTs had with antiquated information technology. Table 9.1 lists this issue, along with two of the many solutions the focus groups commonly raised: "Install newer film readers and a real-time computerized log book." The first suggestion was acted on promptly, and film readers began to be replaced in USAirways line stations as early as November 1993.

The suggestion for a computerized log book took longer, eventually realized not through MRM, but through another change program. During a technical "reengineering" effort in July 1995, the idea of a direct data entry (DDE) feature to improve AMTs' access to the log book surfaced. The technical specialists charged with developing the DDE were unaware of the previous computerized system (Piedmont Airline's log book system acquired during one of the USAirways mergers), or of the recommendation for its reintroduction during the MRM, and thus created the DDE from scratch. Needless to say, it would have been more efficient had their efforts been combined, but the DDE computerized log book was finally put into use in March 1996. It has so far been very well received by AMTs. Despite the extra effort, AMTs are more than willing to enter maintenance information into the computer so that others in the system can access it immediately. They also continue to use hard-copy log books, which must remain on the airplane.

Patterns of Paperwork Errors

Figure 9.1 graphs USAirways maintenance log book errors and maintenance paperwork errors for 1991–1997; all available data during that period are shown. The data series from August 1991 through April 1992 represents the period before the MRM research began. After a 14-month lapse in data collection, the data series resumes in August 1993 and continues (with minor lapses) through July 1997. Various MRM-related events are shown on the bottom of the chart, and other company initiatives affecting paperwork errors are arrayed across the top.

Figure 9.1 reveals a pattern of increasing errors until the MRM project began. Not much effect on the system would be expected during the period of the focus groups (May 1992 to April 1993). Following the MRM paperwork training during the summer of 1993, the rates diminish for about one year (until August 1994). The MRM Research Team felt they were achieving success as measured by the error trends and also by less-quantitative indicators. Log book errors held at a stable level following the mechanics' rewrite in 1994, and they declined further to a new steady state shortly after the computerized log book (DDE) was introduced in early 1996. Total paperwork errors declined steadily to new lows during 1995 and 1996, but 1997 showed a marked increase.

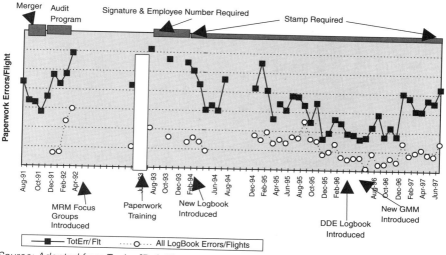

Fig. 9.1 Six years of paperwork errors for all line maintenance stations.

PHASE THREE: Broadening MRM's Scope from Paperwork Errors to All Quality

MRM "Roundtables"—Learning from Specific Errors

In many reported safety incidents throughout the industry, technical procedures and their importance were transmitted by technical bulletins (which in some cases required readers to initial a roster when they had read the document). Throughout the industry, AMTs are often provided technical information for rote memory, and not for consideration in the larger context of aircraft safety and system operation. But technical tasks often require critical thinking. Roundtable discussions at USAirways have proved to be an effective medium for that kind of communication.

The MRM Research Team was encouraged by the error reduction achieved by the MRM program after its first 18 months. They were also impressed by the increasing willingness of the AMTs and foremen to volunteer information about problems and their interest in helping to fix those problems. At that point, the company, the union, and the FAA agreed to institute constructive problem-solving meetings about *specific* safety incidents so that involved AMTs could help the MRM Research Team understand contributing factors, probable causes, and possible solutions.

USAirways, the FSDO, and IAM felt it was important to understand reasons behind specific errors, not only to prevent such errors from recurring, but also to prevent errors in general. The resulting USAirways maintenance problem-solving sessions soon became known as "roundtable" meetings.

The roundtable meetings were open, constructive, and effectively nonpunitive, although amnesty was not promised (and not always delivered). Roundtable meetings were usually limited to the erring parties and the MRM facilitators from the IAM, the FSDO, and QA.

The Experience of USAirways

Between 1994 and 1996, some 20 roundtables were conducted [Ref. 4; 5; 6]. Following most of these sessions, a short article in the USAirways maintenance department newsletter (*Faces & Places*) was written by the erring parties describing the results of the meeting and lessons learned from the discussion. *Faces & Places* is freely distributed throughout the maintenance system and is widely read. The cases reviewed in these sessions covered a wide range of issues, and specific solutions often evolved from the discussions. Not all of the incidents involved paperwork errors, of course, but several of the roundtables did improve work documentation [Ref. 4]. Few roundtable discussions were held during 1997, and this may be related to the 1997 total paperwork error rate increase shown in Figure 9.1.

PHASE FOUR: Broadening MRM to Become a Culture

Training in Communication and Safety

USAirways gained valuable experience with their MRM program. During 1994–1995, the senior vice president implemented other changes to maintenance jobs, organizational structure, and work processes (mostly embodied in self-directed maintenance work teams and reengineering). As these changes began to take hold, the company, the union, and the FSDO decided to endorse a next (and larger) step for MRM—that of changing the very culture of the maintenance organization to achieve more open and trustful communication. The emphasis was on improving safety by reducing errors.

The MRM Research Team realized the advantages of the bottom-up focus of their original efforts and decided that cultural change would require some sort of training program. Such training, they reasoned, should be jointly developed by AMTs and management with subject matter experts (SMEs) providing advice as needed. The development of this training was undertaken in 1996 and completed during 1997.

Several types of SMEs were involved, including developers of training curricula, AMTs with experience in team training techniques, and USAirways pilots and flight attendants with experience in cabin and cockpit resource management programs. SMEs mainly provided information in communication training content (e.g., active listening skills, consensus decision making, assertive communication) and advice on curriculum development (e.g., making and using training videos, writing facilitator notes, preparing workshop manuals). They were invited in on an "as needed and as available" basis.

Little information on the technical aspects of maintenance was sought during this development. The USAirways maintenance technical training department was not much involved, nor were representatives from USAirways' ongoing maintenance reengineering or self-managed teams efforts. They should have been. They undoubtedly will be included in the future.

Ongoing review of the training development was sought from USAirways' AMTs, their union, their maintenance management, the FAA FSDO, and from others external to the company. This MRM development process has resulted in a two-phase, 16-hour training course in human factors knowledge, safety awareness, and communication skills. *All* maintenance and technical operations personnel will undergo this training, as will airworthiness inspectors from the FSDO. The company will continue to evaluate the effects of the MRM intervention using attitude, behavior, and performance measures.

A Parting Thought: The Bases of Success

Management Support

The USAirways case exemplifies positive management support of MRM. The maintenance senior vice president, when approached by the IAM and FSDO with the idea of MRM as a maintenance variation of the successful "altitude awareness program," enthusiastically supported the involvement of several line maintenance stations and provided the resources for collecting paperwork performance data and for mounting an extensive paperwork training program. Twelve months later, his successor introduced some changes of his own, such as self-directed teams for maintenance; he was even more enthusiastic about MRM and encouraged his staff to support it.

When presented with the 1993 MRM focus group recommendations, the senior vice president willingly sponsored implementation of the suggested ideas, at least on an experimental basis. Throughout this process, IAM local leadership and FSDO management were extremely supportive of the MRM initiative. This tripartite cooperation was responsible for the current participation of AMTs and management in the design of the MRM training program for all maintenance people.

Quality of the Intervention

The original initiative was clearly relevant to the AMTs and foremen that it touched. The MRM results show that observations and suggestions made by line maintenance foremen and mechanics were appropriate and useful, because they led to the intended reduction in errors in log books as well as in paperwork overall [Ref. 3].

The examination of the time-series paperwork discrepancy data before, during, and after several MRM interventions (Fig. 9.1) demonstrates that those interventions could be examined separately from the effects of a number of coincident but independent company-initiated changes in documentation procedures. These coincidental interventions included using the "crackdown" technique of more stringent paperwork audits, or "foolproofing" the sign-off process with individual pre-inked stamps.

The form for the aircraft log was designed by mechanics in one line station over a several-month period in late 1993. Log book error data and interviews with AMT users confirmed the success of the new log book. Seven line maintenance groups at two stations interviewed during June 1994 said they preferred the new log book over the old one.

Since the conclusion of the initial study in 1994, the notion of foremen and mechanics getting involved in improving the forms they use has quickly spread and is becoming an idea "in good currency." For example, in 1994, several of the company's line maintenance stations (not among those involved in the original MRM data collection or experimental phases) demonstrated their interest in participating in forms design. Foremen from one of those stations offered to help redesign their "A" Check forms for several aircraft models. Mechanics and leads at another station independently developed a new "A" Check form for the Boeing 757 fleet of aircraft they service.

This ground swell of interest is reflected in the number of AMTs throughout the company who also participated in the review and updating of their general maintenance manual and administrative manual. The new maintenance policies and procedures manual (MPP) is the direct result of the MRM project [Ref. 5], because earlier focus groups often cited the existing manuals as a source of subsequent paperwork errors. The response involved over 100 people, who reviewed and rewrote their core maintenance documents. After about six months of work, the rewritten sections were validated for accuracy by the appropriate departments. Management reported great satisfaction with the quality of the review and rewrite work and the "user-friendliness" of the new MPP—completed largely by AMTs [Ref. 4].

Feedback for Reinforcement

In 1993, results of the initial problem and solution-gathering focus groups were reported to company management, the union, and the FSDO. That report motivated management to implement a hotline and provide paperwork training for all line maintenance employees. In the case of the new form designs, the quality and relevance of the intervention created its own momentum.

The news of mechanic involvement in the aircraft log book redesign traveled fast, and enthusiasm on the shop floor was apparent. The process of continuing to collect and report outcome data during 1995–1997 required a degree of ongoing management support that was difficult to achieve. Nevertheless, self-regulating systems require continuous feedback.

In 1996, the senior vice president endorsed a joint labor-management safety process and announced that MRM was the vehicle to deliver it. With this renewal of management commitment, MRM at USAirways has become a force for cultural change, not merely a passing idea. If you're not already doing it, look again at USAirways' MRM program. It works.

References

[1] Lewin, K. "Action research and minority problems." *Journal of Social Issues.* (vol. 2, 1946): 34–36.

[2] Elden, M., and R.F. Chisholm. "Emerging varieties of action research." *Human Relations.* (vol. 46, 1993): 121–141.

[3] Taylor, J.C. *Maintenance Resource Management (MRM) in Commercial Aviation: Reducing Errors in Aircraft Maintenance Documentation* (Final Report of 1993–1994 Project Work). Sociotechnical Design Consultants. Box 163, Pacific Palisades, CA 90272, December 1994.

[4] Kania, J.R. "Panel presentation on airline maintenance human factors." *Proceedings of the 10th Meeting on Human Factors Issues in Aircraft Maintenance and Inspection.* Washington D.C.: FAA, Office of Aviation Medicine (1996).

[5] Driscoll, D.D. "Maintenance resource management." *Faces & Places.* Pittsburgh: USAirways Maintenance Operations (August 1996): 5–10.

[6] Marx, D.A. "Learning from our mistakes: A review of maintenance error investigation and analysis systems." In *Human Factors Issues in Aircraft Maintenance and Inspection '98* (CDROM). Washington, D.C.: FAA, Office of Aviation Medicine (http://www.hfskyway.com).

# Chapter 10

Early Successes with Open Communication

OVERVIEW: Continental Airlines' MRM Communication Program Pays Off

By 1988, airline maintenance managers were beginning to think seriously about the importance of improving communication, teamwork, and participative decision making. A program to improve communication in the cockpit, Cockpit Resource Management (CRM), had already proven effective, even under some of the most harrowing circumstances [Ref. 1]. CRM's success arose from a recognition that the aviation work system is more than individuals operating separately, that severe stresses impinge on effective decision making, and that information about system purpose and system performance is necessary for improved communication practices, better teamwork, and better stress management. Continental Airlines put these insights to work with its 1991 introduction of an MRM program that borrowed key ideas from the Cockpit Resource Management program. In communication and in several bottom-line measures, the payoff was impressive. Continental scored significant gains over a number of years until management shifted its primary focus to cost reduction, other gains slowed or reversed.

Management Culture in Aviation

To reduce the chance of human error, management has three options. Changes can be made in the worker, the work, or the context of the work (its technology or organizational systems). To its detriment, aviation maintenance has focused considerably more energy on changing the individual worker and the work than on changing the tools, or the system. As late as 1990, observers from the social and behavioral sciences continued to find that the culture of the commercial aviation industry overly emphasized the individual, while largely ignoring the critical importance of teamwork [Ref. 2 p. 495].

Sources of Stress in Maintenance

The lack of team support in maintenance causes unnecessary stress on individual AMTs. Tight schedules and fixed deadlines are also highly stressful facts of life in commercial aviation today. Management-imposed deadlines are often seen by AMTs as arbitrary, unreasonable, or unattainable. Management emphasis on meeting schedules increases the pressure on AMTs to compromise [Ref. 3]. Often, mechanics and their managers are faced with a Speed-or-Accuracy Trade-Off (SATO) [Ref. 4]. To deal with the SATO dilemma, management may choose to delay departure, transfer work elsewhere, or defer repairs. To choose any one of these alternatives, however, is to raise feelings of helplessness, frustration, and misunderstanding among AMTs and their foremen. This frustration is stressful, sometimes leading to physical as well as psychological strain [Ref. 5]. Foremen and AMTs seldom have the information and the means to control deadlines. This lack of decision-making latitude has been shown to increase stress and the likelihood of errors [Ref. 6]. The growth in professionalism required of AMTs today is stunted by their lack of control or effective participation in setting deadlines.

From research conducted in 1989–1990, it was shown that AMTs in heavy maintenance believed that their foremen and managers have both the *means* of control and the information required to do so [Ref. 7]. The AMTs saw themselves as largely powerless to control the source of deadline stress. Better collaboration and shared decision making among the AMTs and their management seems like a natural solution to this problem.

Teamwork in Maintenance

Research conducted in heavy maintenance hangars concluded that the sites with the best communication patterns and highest productivity had managers who encouraged crew briefings, supported cross-group communication, and emphasized mission-centered performance objectives. At other job sites, the 1989–1990 study echoed earlier reports that the overall culture of the industry continued to overemphasize individual work over group work, reactive rather than proactive response, and punishment rather than reward [Ref. 7]. To overcome this cultural resistance to teamwork and participation, the study concluded that the teamwork and communication training program successfully used with cockpit crews during the 1970s and 1980s (CRM) should be adapted for use in aviation maintenance.

Let's review what CRM is all about, and then we'll move into the maintenance version of this approach.

CRM and Teamwork Training in Aviation

The Cockpit Resource Management training approach was introduced into flight operations during the late 1970s, following several aircraft accidents that clearly involved poor flight crew communication [Ref. 8]. It then spread from civil to military aviation and into air traffic control, cabin service, and maintenance. CRM involves training in several team-related concepts: communication skills, self-knowledge, situation awareness, and assertiveness skills.

CRM training improves system coordination, quality, and safety. While safety may be the central focus of the training, Maurino and his colleagues have concluded that safety performance is a result of many organizational and management factors, not just a few. Safety training helps to determine the quality of the *overall* system, and that in turn delivers a wholesale impact on safety performance [Ref. 9].

An adaptation of CRM to maintenance operations was reported soon after the 1988 Aloha accident [Ref. 10], but it pursued very limited objectives and achieved only limited success. Shortly thereafter one airline company, Continental, achieved considerable success in training that used a modified version of CRM training for all maintenance management and technical staff support people, as well as a sprinkling of AMTs. Their program succeeded in improving attitudes and in boosting maintenance safety results [Ref. 11; 12; 13]. Following a several-year lapse, Continental has continued to extend and modify its program. It now targets its AMT workforce more specifically than before, and it calls its program MRM.

Continental Airlines' pioneering effort to improve its maintenance communication has been an inspiration to numerous companies in the airline industry. Its successes deserve to be emulated.

Continental Introduces MRM

In 1991, Continental Airlines launched its new program for improving maintenance communication. The idea was the product of discussions between the senior vice president of technical operations and the director of Continental's human factors and cockpit resource management program. The two of them agreed that the company's success in its CRM flight deck communication program could be adapted for maintenance, starting with improving its management communication skills first. A training course was developed by maintenance quality assurance and maintenance operations managers, with help from subject matter experts (SMEs) from Continental's maintenance training section, trainers from the company's human factors/cockpit resource management program, and university-based training evaluation specialists.

Management Support

The program had strong support at the highest maintenance level. This training, originally called Crew Coordination Concepts (CCC), was the personal agenda of the senior vice president. He intended it for his entire department's management—and eventually the whole workforce of AMTs. Following the company's model of cockpit resource management, this training was designed to improve safety and efficiency by identifying and modifying negative behavioral norms about safety, by using assertive behaviors effectively, by understanding individual leadership styles, by managing stress while improving problem solving/decision making, and by enhancing interpersonal communication skills. Maintenance operations people conducted the training with assistance from professional training consultants. Each two-day training session began with a reference to the vice president's direct interest and confidence in the program. Sometimes senior managers personally introduced the program to participants on behalf of the senior vice president.

The training that followed included a combination of short lectures, discussions in large and small groups, individual study, and role-playing maintenance cases. The training was accomplished by rotating the location of the program among three hub cities and repeated weekly with groups of about 20 participants each.

The program's original goals were achieved. By all measures, CCC was a successful communication training program. At the end of two years, all 1,800 management and professional staff had completed the training. Thereafter, and before the training was reduced to a trickle, some 300 quality inspectors and mechanics also had attended the course over a ten-month period. Enthusiasm for the course was high. As managers and staff professionals finished their training, over 90% of them reported that it would benefit themselves and others, and that it would substantially change their behaviors on the job [Ref. 12]. Later, as the AMTs attended their sessions, more than 80% of them also said the training would improve safety and teamwork [Ref. 13]. Even more importantly, in the nearly three years the program existed, it proved successful in reducing accidents, incidents, and injuries in maintenance operations.

Management Attention Correlates with Program Success

Despite its successes, the program lapsed when top management attention was diverted to other matters. Before the end of that three years, the senior vice president had turned his attention to cost reduction and efficiency. Not long after that, he left the company. His replacement, too, had other priorities to attend to before management could once again focus on MRM.

MRM and CRM Compared

The experience at Continental shows that MRM shares with CRM the basic issues of communication and team coordination and an interest in evaluating resulting attitudes and behaviors that relate to those issues. But MRM, as applied in Continental's CCC program, differs from CRM in several important ways: (1) The maintenance target audience is more diverse than are cockpit crews. (2) The maintenance program includes AMTs, staff support personnel, and management. (3) Maintenance activities are typically carried out over a longer period of time than flight operations. (4) Although line maintenance does experience the sometimes intense "one flight at a time" environment that is typical of the cockpit, base maintenance and repair usually have longer time horizons, and they must meet criteria other than just "flight pushed back on time." MRM goals include the reliability of technical operations processes, occupational safety, and airworthy aircraft, as well as the reduction of aircraft accidents. Another important difference between MRM and CRM evaluations is that MRM programs draw on a wide variety of objective performance data to test outcomes. From the beginning, MRM was intended to impact maintenance error rates—it was created to improve human reliability in measurable terms [Ref. 14]. Early versions of cockpit resource management programs were typically assessed in terms of accidents prevented, while MRM programs are more readily assessed in terms of performance achieved.

MRM Performance Measures

A number of robust indicators of maintenance performance were used at Continental. Data were collected monthly for a substantial period prior to and following the onset of the airline's MRM training. The key outcomes measured were aircraft ground damage, occupational injuries, and maintenance delays. The performance shown by these indicators before and after the training were first published in the report of a NASA-sponsored longitudinal study [Ref. 12]. Posttraining performance improved on all three indicators, and it continued to improve for at least as long as management attention remained focused on MRM.

MRM Results at Continental

Safety improved—occupational injuries had been increasing in the six months prior to training, but the number of injuries began to decline shortly after the training began. Figure 10.1 displays those results.

The pretraining and posttraining data are reported over the 30 months between January 1991 and June 1993. Fig. 10.1 also shows the effect of the change in management strategy after June 1993. The two spikes in injuries in September 1993 and January 1994 are each separately accounted for by a single maintenance base station, and each spike occurs during the month the closure of each base was announced. Those dramatic increases in lost-time injuries probably reflect a decline in morale as AMTs learned about the

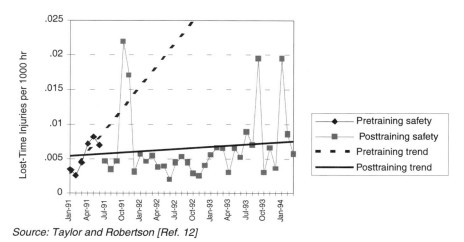

Source: Taylor and Robertson [Ref. 12]

Fig. 10.1 Occupational injuries before and after MRM training.

closures. Once management turned from communication and safety to plant closures and cost cutting, the excellent results of their MRM program reversed.

Figure 10.2 graphs ground-damage incidents during the same period. Although maintenance-related ground damage was trending downward before the training began, the rate improved until August 1993, showing a drop of nearly 60% in the rate of damage reduction. Ground-damage data were not available for three months after August, but as Fig. 10.2 shows, the number of incidents increased again in December 1993 and January 1994. Again, so long as management attention remained on MRM objectives, performance against those objectives improved.

Dependability Improves with Communication Training

Dependability—that is, on-time performance—continued to improve after the MRM training. On-time departures showed improvement before the training began, and the trend continued for the period data were available following the training, as shown in Fig. 10.3. Although the culture shift accompanying MRM training does not exhibit a further positive impact on this non-safety related measure, Fig. 10.3 also shows that the quest for a safety culture also does not diminish productivity either.

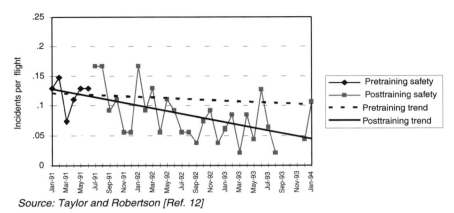

Source: Taylor and Robertson [Ref. 12]

Fig. 10.2 Ground-damage performance before and after MRM.

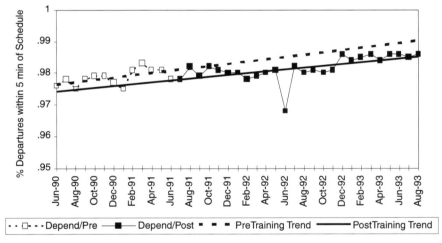

Source: Taylor and Robertson [Ref. 12]

Fig. 10.3 Dependability before and after MRM.

MRM Attitude Measures

Attitudes, perceptions, and intentions related to MRM training were evaluated at Continental through survey questionnaires. As with CRM, MRM programs have a strong interest in evaluating results.

The evaluation of the program's effectiveness includes measures of participant attitudes and opinions similar to those developed by Robert Helmreich and his colleagues at the University of Texas [Ref. 10; 15]. The resulting MRM questionnaire, called the Maintenance Resource Management Technical Operations Questionnaire (MRM/TOQ), includes measures adapted specifically for people employed in aviation maintenance. The MRM/TOQ was developed, tested, and proven during the three-year evolution of Continental's CCC/MRM training program [Ref. 16].

MRM Training and Changes in Attitudes

Comparing managers' attitudes immediately after training with their pretraining attitudes showed significant improvement on three scales [Ref. 12]. Statistically significant improvement took place in attitudes toward "usefulness of communication and coordination," and "recognition that stressors affect decision making." Attitudes toward "willingness to share command responsibility" also improved significantly following the training sessions. The same three attitudes of these managers two, six, and twelve months later reveal that their favorable posttraining attitudes remained at those positive levels for many months after the training. The fourth attitude scale measured "willingness to voice disagreement" (a central element of assertive behavior). Although this measure showed no significant change immediately following training, results improved significantly two months after the training and they remained at that higher level six and twelve months afterward. Figure 10.4 graphs the managers' attitudes over those five periods. The attitude scale ranges from "1" (strongly disagree) to "5" (strongly agree) for each attitude.

As Fig. 10.4 shows, the expected influence of the training on all four attitudes was found to be stable and strong. The improvements showed more than merely a brief "honeymoon effect" of good feelings immediately after the training.

Figure 10.5 uses the same attitude scale values as Fig. 10.4. Figure 10.5 compares the managers' pre- and posttraining attitudes with those of the AMTs who attended the training in the months just before the program was suspended. This comparison is incomplete because the AMTs were not surveyed in the months following their training. In the main, however, the AMTs' attitudes parallel those of their managers. Of the four attitude scales depicted in Fig. 10.5, only one, the importance of managing stress, is substantially lower for the AMTs; however, they nevertheless showed a significant strengthening in this attitude. Like the managers, AMTs' attitudes toward assertiveness did not improve on the first posttraining survey, but, unlike the managers, AMTs were not resurveyed in the months afterwards and thus cannot be tested for the delayed effect that was shown by the managers.

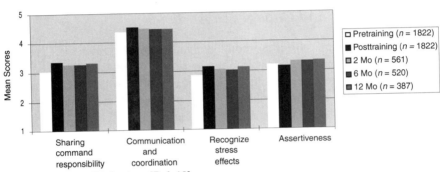

Source: Taylor and Robertson [Ref. 12]

Fig. 10.4 Maintenance managers' attitudes before and after MRM training.

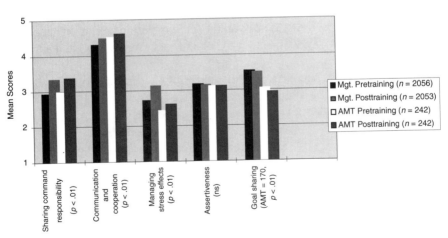

Source: Adapted from Taylor et al. [Ref. 13]

Fig. 10.5 Pre- and Posttraining attitudes of AMTs and maintenance managers.

Importance of Goal-Sharing Control Variable

The validity of the MRM/TOQ survey was itself tested by including items that were *not* expected to change as a result of Continental's MRM training. Although the work groups' goal setting and goal attainment activities are important aspects of maintenance work, the MRM training was not designed to directly influence their attitudes or perceptions about that aspect of management. Figure 10.5 thus includes pre- and posttraining comparisons of opinions about goal sharing for both the AMT sample and the maintenance management sample. As expected, if the training has no impact on this opinion, the bar chart reveals no significant change in management or AMT opinions about goal-sharing and goal-attainment. This early version of MRM training did not positively influence participants' views of goal sharing—nor was it designed or intended to do so. Compared with the managers, AMTs held a significantly lower opinion on goal sharing. These results reinforce the observation made earlier in chapter 5, and elsewhere, that AMTs are historically less involved in the important "big picture" issues than their managers are. This also reminds us—as we saw in the study about "maintenance speedometers" reported in chapter 8—that communicating goals and outcomes on the maintenance floor can result in powerful improvements in performance.

Relationships Between Attitudes and Performance

Continental also examined whether changed attitudes following the training could be associated directly with the improved safety and other performance indicators that followed the training. The short answer is yes.

Relationships between the four attitude scales and maintenance bottom-line performance were analyzed over many months. This time-lagged or longitudinal correlation gauged the effects of the training program on work performance. The analysis was conducted first for the managers [Ref. 12; 17] and later for the AMTs [Ref. 13].

Management Results

Figure 10.6 graphs statistically significant positive correlations between the four posttraining attitude scales and subsequent on-time departures and occupational injuries. Ground-damage numbers are not included in Fig. 10.6

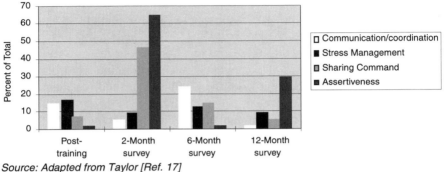

Fig. 10.6 Management MRM attitudes and performance.

because few of the management attitudes were found to be directly correlated with it in any of the time periods measured. The remaining two performance indicators were measured periodically for the 24 months following the MRM training. These two measures together form a combined total of 48 correlations between attitudes and performance. The bars represent the percentage of positive correlations, which were large enough to exceed preset levels of significance. The training did indeed have a positive effect on on-time departures and occupational injuries. The possibility of an on-going cultural change became real, requiring only a management commitment to make the MRM way of doing business a constant fact of life in maintenance work.

Figure 10.6 shows that most attitude scales exceed 5% for the time periods, supporting the prediction that MRM training should have some effect on subsequent performance. But there are some spectacular periods in the chart, especially from the survey taken two months after the training. In particular, the remarkably positive correlations between assertiveness and sharing command, and the performance outcomes indicate that the lessons learned from MRM training coalesce and strengthen in the months afterward, and produce impressive performance gains.

The best results in Fig. 10.6 occur at the two-month mark following the training. Thereafter, positive correlations trended downward. Thus, very positive effects of training on managers' attitudes appear to diminish over time. Even then, the relatively strong, positive influence of the value of assertiveness (30% of all correlations are significant) still held by individuals twelve months after their training marks an important outcome and emphasizes how long-lived the effects of the training can be.

AMT Results

The three performance indicators included aircraft ground damage, days lost to occupational injury, and on-time departures. These performance indicators were tracked from August 1993 through January 1994 (six months after the AMT's MRM training began). Because a much smaller number of months had elapsed following the AMT training, a different analytical technique was used for the AMT sample. A six-month average score was calculated for each of the three outcome measures for each work unit, and this six-month performance average was then correlated with the average AMT attitudes for those same work units. The results are as remarkable, not only because of strong correlations, but also because they showed a complementary effect to the earlier management analysis [Ref. 13].

Two of the posttraining attitude scales, "recognition of stress or effects" and "assertiveness," were strongly related to the number of ground-damage incidents. These relationships imply that the more assertive the AMTs are about safety, and the better they recognize and manage their own stress, the fewer the incidents of ground damage. The second safety indicator, "occupational injury," was not found to be significantly related to any of the AMTs' posttraining attitude scores. The third measure, "dependability," showed a very strong relationship with attitudes about "sharing command responsibility." Good on-time performance was thus found in line maintenance stations with AMTs who held favorable attitudes toward participation and sharing command responsibility.

Managers' Opinions about Goal Sharing Related to Performance

As shown in Fig. 10.5, the control variable "goal sharing" behaved as expected and did not change or improve for managers following their MRM training. But communication about goal setting and goal attainment is important in achieving maintenance performance irrespective of MRM training. Goals are the "what" of communication, just as assertiveness and participation are the "how" of communication. And "what" and "how" are equally important in effective workplace communication.

Analyses of the correlations between opinions toward goal sharing and the outcome measures have recently been completed. Results show that opinions toward goal sharing were significantly related to on-time departures over nearly two years. But goal sharing was found to be much less related to the other outcomes—aircraft ground damage and lost-time injuries. Figure 10.7 shows significant correlations between Continental maintenance managers' assessment of goal sharing in maintenance and its effect on on-time departures. These results are impressive. They suggest that on-time departures are indeed a major topic of daily communication among at least some maintenance managers.

The absence of such strong associations with other outcome measures, such as ground damage and personal safety, does not mean that those criteria were not discussed among managers, but it does suggest that they were not as common or consistent a topic. In comparison, the AMTs' results showed

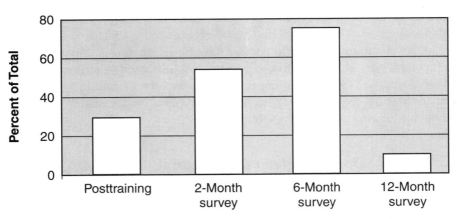

Fig. 10.7 Correlations between dependability (on-time departures) and management opinions toward goal sharing.

no significant correlations between their posttraining opinions about goal sharing and any subsequent outcome measures. We may conclude from this that some managers were involved in frequent communication about on-time departure goals and that this helped them direct their AMTs to improve or maintain high goal attainment. But the AMTs themselves were not much included in goal-sharing communication.

Success and Management Support for MRM

Successful change requires the unequivocal support of top management, and sustained success requires continued top management support. Continental's senior technical operations vice president created a program for all 2000 of his management and staff support personnel. The case at Continental shows that if executive commitment lasts long enough, and if its sponsorship is persistent and visible, then positive results continue. Maintenance managers at Continental had begun to seriously value the open, assertive communication, the safe work habits, and the problem-solving methods that MRM programs espouse. Those managers reported in surveys and interviews that this program, unlike most others they had participated in, really worked. They believed that top management's support was genuine. They hoped that it would continue and that the program could be sustained through a refresher course to further develop and practice their newly acquired skills [Ref. 12 pp. 49–50]. Unfortunately, the dream halted there. Although it was not officially terminated, the recurrent MRM training for managers and the first-time training for AMTs proceeded only haltingly.

Success and Quality Intervention

Successful change requires a well-conceived and relevant intervention in the existing work culture. Continental's program was well planned and effectively executed. It was based on the best of the lessons learned from flight-deck CRM training. The training was managed, administered, and conducted by technical operations line managers who, not incidentally, also had good training skills. The training content and its illustrations were relevant to the work the participants do. In addition, the training required participants to engage in active, student-centered learning. As a result, participants acclaimed the training as unparalleled in appropriate content, timely in the delivery of its message, and useful in its ability to be practically applied. As increasing numbers of people were trained and its effects became manifest, most participants recommended recurrent training, and many endorsed the wider use of the training throughout the company. This positive experience laid the groundwork for future success.

Success and Feedback

Successful change requires timely, appropriate feedback. The senior vice president believed that the training program would improve human resource management [Ref. 14]. The first day of each training session opened with a statement of purpose for the program: "To equip all technical operations personnel with the skill to use all resources to improve safety and efficiency." In this case, FAA and NASA funding was available to facilitate and enable novel and extensive data collection before and during the period of MRM training. Such support permitted appropriate analysis of those data to test the effects of the training on airworthiness, on-time departure, and personal safety. The data collection, analysis, and reporting were done with strict confidentiality by a neutral third party, ensuring honest responses to the surveys and interviews.

Given that those measures of MRM-relevant attitudes and behaviors were reliable and valid [Ref. 16], the results provided encouragement to continue the effort. The results themselves were quite positive from the start [Ref. 11]. Although all reports were available to all participants, they were not widely distributed or publicized within the maintenance department. A final report to NASA traced attitudes, behaviors, and performance results for over three years [Ref. 12]. While the results were positive, they revealed a diminishing relationship between attitudes and performance after the first year. The MRM program arrived at Continental without fanfare in 1991. In 1994, it was suspended without fanfare. How the program was intended to relate to the overall mission of the company, or to the maintenance mission, was never explicitly communicated to either management or to the workers.

The Importance of Purposeful Systems

Most airline managements don't communicate much about the reasons AMTs or foremen do what they do. Even when companies do communicate with AMTs, there is not much evidence of that AMTs communicate in turn about the what, the when, the how, or the why of the work they do. This lack of mutual, purpose-centered communication is, by all signs, driven by the assumptions and expectations of managers and staff specialists as much as it is by the AMTs themselves.

It's important to communicate "why," to offer explanations to mechanics, inspectors, planners, and their supervisors. Not all AMTs want more communication, but many (especially the younger ones) do (see Chapter 5). There is some truth in the assertion that American industry (including airlines) provides too much information to employees and not enough communication. What is discussed is not often what is most important, nor is the communication carried out in the most effective way. This issue is not one of the quantity of information, but of its quality—and whether or not it should be discussed.

Communicating Purpose

Maintenance managers must be clear about the core purpose of their company—its distinctive competence. And they must be clear about the special purpose of the maintenance function and its relation to the company's larger purpose. Field observations between 1989 and the present reveal that the maintenance purpose is communicated poorly to maintenance foremen and rarely, if at all, to AMTs. In a few companies, the maintenance purpose is written down and posted where AMTs can see it. In most companies, however, not even the foremen can recite or define the central purpose of their maintenance work, or the company's purpose, or its distinctive competence in the air transport industry. AMTs are further hampered by the lack of a clearly articulated sense of shared purpose. Long-term objectives are poorly communicated. If the maintenance purpose and its key objectives are not communicated, there is little reason to expect AMTs to focus their energies on achieving them. Short-range goals, such as flight departure times, or the expected time for an "A" Check, *are* known to AMTs, but in many companies the expectations to meet such goals are imposed from above and they are unrealistic. When more realistic short-term goals are developed in cooperation with AMTs, the guiding attitudes and the near-term performance of a maintenance station usually improves.

The MRM/TOQ survey used at Continental during 1991–1994 included the set of questions exploring respondents' perceptions of how work goals are communicated and discussed within and between work groups. These questions were included in that questionnaire for two reasons. First, these questions were included as a test of the "halo effect" in survey research, an unwanted event where answers to *all* questions improve following a positive intervention such as the MRM training. As noted above, that unwanted effect was not found. Second (and more important for our purposes in this chapter) these questions were included because the concepts they measure are acknowledged to be related to organizational effectiveness in other industries [Ref. 18; 19]. It is important to see how they are similarly related to outcome performance in airline maintenance. As shown in Fig. 10.7, line manager's opinions about goal-sharing are strongly related to their work unit's on-time performance. On the other hand safety was not found to be consistently related to goal-sharing. Obviously, "no delayed flights" is an important goal for every airline. But what if safety goals *were* communicated and shared as a central part of MRM training and daily on-the-job practice? We would expect direct and dramatic results in that all-important area too.

The structure of maintenance organization should facilitate purposeful communication and cooperation rather than heighten differences and increase rivalry and conflict. The term "functional silos" [Ref. 20] refers to an organization and communication system too often seen between different departments. Rather than working out conflicts at the source, troublesome issues are sent up their separate "silos" for resolution at the top with command decisions coming back down the silos. Such a once- or twice-removed approach to workplace communication creates slower and less than fully effective problem solving. This style of communication has often been observed between maintenance and other departments too: material services, inspection, customer service, planning, and dispatch. Specific results from Continental's MRM study suggest that goal sharing with other departments was reported mainly by top maintenance management [Ref. 12 pp. 12–13]. For middle and lower management, staff support specialists, and AMTs, goal sharing between functional silos before the training began was much lower. Such MRM training can be explicitly designed to help overcome the troublesome effects of the silo structure.

By design, Continental's training often included participants from various departments. This simple expedient, however, was not enough to change behaviors and "us versus them" attitudes that are created in silo organizational structures. A change in training emphasis, coupled with changes in work design that includes multifunction communication, could mightily improve performance objectives and overall goal attainment.

A Parting Thought: Borrow from the Best

The MRM training experience at Continental set an industry standard for improving attitudes and operational performance through training in open communication. It proved that such training is enthusiastically embraced by maintenance participants. Assertiveness and participative management practices are thus powerful tools that improve maintenance effectiveness. Continental's experience also demonstrates the importance of active and visible top management support, along with the regressive effects that occur when that support is reduced or withdrawn. Finally, this case suggests that MRM programs can and should be further improved to enhance goal sharing between departments as well as within them. Continental's approach is well worth emulating.

References

[1] Helmreich, R.L. "Managing human error in aviation." *Scientific American* (May 1997): 62–67.

[2] Hackman, J.R. *Groups That Work.* San Francisco: Jossey-Bass, 1990.

[3] Bruggink, G.M. "Uncovering the policy factor in accidents." *Air Line Pilot* (May 1985): 22–25.

[4] Drury, C.G., and A. Gramopadhye. "Speed and accuracy in aircraft inspection." *Position Paper for FAA Biomedical & Behavioral Sciences Division.* Washington, D.C.: Office of Aviation Medicine, 1991.

[5] French, Jr., J.R.P., R.D. Caplan, and R.V. Harrison. *The Mechanisms of Job Stress and Strain.* New York: Wiley, 1982.

[6] Karasek, R., and T. Theorell. *Healthy Work.* New York: Basic Books, 1990.

[7] Taylor, J.C. "Maintenance organization." In W. Shepherd et al. (eds.). *Human Factors in Aviation Maintenance Phase 1: Progress Report.* Washington, D.C.: FAA, Office of Aviation Medicine, 1991.

[8] Wiener, E.L., B.G. Kanki, and R.L. Helmreich. *Cockpit Resource Management.* San Diego, CA: Academic Press, 1993.

[9] Maurino, D.E., J. Reason, N. Johnston, and R.B. Lee. *Beyond Aviation Human Factors.* Aldershot, Hampshire: Ashgate Pubs., 1995.

[10] Taggart, W. "Introducing CRM into maintenance training." *Proceedings of the 3rd International Symposium on Human Factors in Aircraft Maintenance and Inspection.* Washington, D.C.: FAA, Office of Aviation Medicine, 1990.

[11] Stelly, Jr., J., and J. Taylor. "Crew coordination concepts for maintenance teams." *Proceedings of the 7th International Symposium on Human Factors in Aircraft Maintenance and Inspection—Science, Technology and Management: A Program Review.* Washington, D.C.: FAA, Office of Aviation Medicine, 1992.

[12] Taylor, J.C., and M.M. Robertson. "The effects of crew resource management (CRM) training in airline maintenance: Results following three years' experience." Washington, D.C.: National Aeronautics and Space Administration, 1995.

[13] Taylor, J.C., M.M. Robertson, and S.W. Choi. "Empirical results of CRM training for aviation maintenance technicians." *Proceedings of the 9th International Symposium on Aviation Psychology.* Columbus, OH: Ohio State University, 1997.

[14] Fotos, C.P. "Continental applies CRM concepts to technical, maintenance corps" and "Training stresses teamwork, self-assessment techniques." *Aviation Week & Space Technology* (August 26, 1991): 32–35.

[15] Gregorich, S.E., R.L. Helmreich, and J.A. Wilhelm. "The structure of cockpit management attitudes." *Journal of Applied Psychology* (vol. 75 1990): 682–690.

[16] Taylor, J.C. "Reliability and validity of the maintenance resources management/ technical operations questionnaire (MRM/TOQ)." *International Journal of Industrial Ergonomics* [Special Issue on Human Factors in Aviation Maintenance], (in Press).

[17] Taylor, J.C., "Effects of communication and participation in aviation maintenance." *Proceedings of the 8th International Symposium on Aviation Psychology.* Columbus, OH: Ohio State University, 1995.

[18] Easterbrook, G. "Driving quality at Ford." In R.M. Kanter et al. (eds.), *The Challenge of Organizational Change.* Macmillan, 1992.

[19] Hostage, G.M. "Quality control in a service business." *Harvard Business Review,* (Sept.–Oct. 1975): 98–118.

[20] The term "functional silos" was coined by G.A. Rummler and A.P. Brache in *Improving Performance* (San Francisco: Jossey-Bass, 1991).

Chapter 11

AMT-Oriented Communication Training Comes of Age

OVERVIEW: The Need for Strategic Thinking and Management Follow-Through

In recent years, airline maintenance departments, large and small, have found encouragement and assistance in undertaking MRM programs. Major trends include

1. Familiarization programs for maintenance managers designed to emphasize the importance of safety and communication
2. An emphasis on specially created MRM training for AMTs

Although these trends are well intentioned and, for most purposes, highly successful, obstacles and pitfalls lie in wait for the unwary. Particular problems include an overemphasis on training AMTs and a resulting underemphasis on sustaining change. One company's experience with MRM training illustrates this effect. This case reveals that problems can be overcome with advance planning and effective follow-through.

Before we get into the case, let's review the effect of earlier MRM training programs.

Background

By the time Continental's program (Chapter 10) was concluded in 1994, its results and case details were well known to others in the industry. To train

2200 people in three years, Continental conducted a session somewhere in the country nearly every week. A stream of visitors observed their training and went away with an idea of what succeeded and how it worked. Those observers included people with maintenance interest or responsibilities from other airlines (foreign and domestic), from the military, from aircraft manufacturers and other vendors, and from government agencies (including FAA, NASA, and Transport Canada).

MRM Training Spreads

The Continental story created a snowball effect as many of Continental's visitors were inspired to do something similar at their own organizations. Since 1994, a number of MRM training programs have been started. Some of these derivative efforts created new course material and curricula. In some cases, whole courses were provided to others, allowing them to take part in the MRM process without incurring extensive development costs. Two of these recent courses are well developed and well publicized. They are the Boeing company's Model Maintenance Safety Program (MMSP) and Transport Canada's Human Performance in Maintenance Program (HPIM).

Boeing's Model Maintenance Safety Program

Boeing's MMSP was created by its safety engineering group in 1994. They designed it in response to the growing international market for commercial aircraft and the increased awareness of the role of maintenance in aviation safety. MMSP was designed to be a brief (four-hour) introductory event and an overview to encourage safety in the worldwide airline industry. As such, it is a management awareness program rather than a training program for specific skill enhancement.

MMSP is organized around several concepts important to maintenance safety performance. Emphasis is on such concepts as the central role of the AMT, the necessity of effective workplace communication, the requirements of error management, and the use of systemic improvement methods. MMSP emphasizes the best maintenance practices for safety. The course material includes the summarized results of Boeing's various studies exploring maintenance safety. Its training material also draws on the results of Continental's

MRM experience—its emphasis on maintenance management awareness of and communication with technicians, and its successful outcomes.

MMSP has been used in a number of Boeing's safety initiatives with airline companies and regulators in Asia, South America, and Africa, and during its first four years, over 6000 people, mainly aviation executives, attended the program. Boeing's commitment to improved management, communication, and safety is manifest in several related programs. First, MMSP has recently been combined with the two-and-a-half day training for Boeing's Maintenance Error Decision Aid program (MEDA), creating a systemic approach to managing for safety and reliability—an approach that emphasizes both safety information and human communication. Second, Boeing has created its Institute of Aviation Leadership, which provides customer airlines, including those from emerging countries, with in-depth (currently 90 days) executive training in management skills. This training is designed to help these executives understand that safety is the most important focus for their rapidly growing companies.

Transport Canada's Human Performance in Maintenance Program

In the early 1990s, Transport Canada developed its own MRM training program, following the 1989 accident at Dryden, Ontario (Chapter 8). Information about safety initiatives and training for maintenance personnel was collected throughout the industry, and this included Continental's MRM training. The resulting Canadian course, called Human Performance in Maintenance (HPIM) was designed specifically for technicians. HPIM bears a family resemblance to Continental's MRM program and its course materials, but it is custom-made for its intended audience. HPIM was first made available as a do-it-yourself training program in early 1994 and soon became widely known because of the relevance of its training materials and its ready availability. Among HPIM's most popular innovations is a set of safety posters called the "Dirty Dozen." A separate poster was created for each of 12 major causes of maintenance errors. Table 11.1 lists the Dirty Dozen.

The HPIM training course was designed to cover, in depth, the first three topics in Table 11.1. Although all twelve items are addressed in the two-day training program, the last three are mentioned only in passing. HPIM's

TABLE 11.1 THE "DIRTY DOZEN"—CAUSES OF MAINTENANCE ERRORS

◆ Lack of communication	◆ Lack of assertiveness
◆ Stress	◆ Lack of resources
◆ Fatigue	◆ Pressure
◆ Complacency	◆ Lack of knowledge
◆ Distraction	◆ Lack of awareness
◆ Lack of teamwork	◆ Norms

course syllabus includes extensive use of video cases followed by group discussion and active learning modules employing group problem-solving exercises. This multimedia program includes ample reminders for participants to be careful in their work and to remember the lessons from the course. The reminders continue even after the close of the formal training sessions.

The two-day HPIM course is available for use free of charge, and the Dirty Dozen posters and related training aids are available at minimal cost. In its first four years, HPIM has been presented, sometimes with local modifications, to many thousands of participants and observers. The major exposure to the HPIM course has been attended by more than 7,000 participants in two airlines in North America and one aviation training/consulting firm. Two Canadian airlines are preparing to make extensive use of HPIM. A follow-up HPIM-2 training program emphasizing "norms" from the dirty dozen, as well as company culture and written communication has recently developed and is currently available through Transport Canada [Ref. 1].

The Increasing Variety of Current MRM Programs

Boeing's MMSP and Transport Canada's HPIM are excellent examples of the range of current developments in MRM training programs. Such programs allow choices of who develops the material and what topics are covered in the training—delivered to whom, by whom, and in what depth. Key features of MMSP and HPIM are found in various combinations in the many emerging MRM programs in North America.

Numerous Topics Covered

The range of material covered by the various new programs is broad, but open, effective communication is always emphasized. Individual programs vary in their use of related topics such as teamwork, group decision making, listening skills, and assertiveness. Other topics found in new MRM training programs include norms of behavior and the development of effective workplace social-system roles; the effects of stress on safety and its effective management; and the impact of the local situation and awareness of it. Recent interest in maintenance error detection, error reporting, and root-cause analysis has also influenced the content of new MRM training packages. Some now include sections on safety data reporting systems and how to create shared safety goals. Examples of this model-mixing trend include USAirways' new MRM program (Chapter 9) and the recent combination of Boeing's MMSP and its MEDA programs.

Form and Structure Vary Widely

Few of the current offerings are copies of Continental's initial model of locally developed, two-day classroom training for maintenance managers. In some cases, material for the new MRM courses is created specifically for AMTs, and sometimes it is intended primarily for maintenance managers. Sometimes training material is merely borrowed and slightly adapted from flight operations' CRM training. Increasingly, the materials from Transport Canada's HPIM course, especially the Dirty Dozen posters, are finding their way into MRM programs of many airlines.

A Single Episode or an Ongoing Process?

Sometimes the new MRM training is intended as a one-time-only event. In practice, such training ranges from a single, four-hour session to two full days. There is increasing awareness that a one-shot training program is not enough. Indeed, the introduction of MRM training signals the beginning of an ongoing cultural change in maintenance. Sometimes the decision for follow-on training is made only after the initial session has been completed. In other cases, the necessity of a continuous process of cultural change is recognized in advance. One MRM program variation that reflects this growing understanding starts with one day of MRM training followed by a pre-planned day of recurrent training a few months later. In the future,

MRM-style training and practice may become fully integrated features of organizational life—artifacts of a new workplace culture rather than add-ons to an older culture that still resists this kind of innovation.

Delivery Varies

Continental's MRM approach made direct use of training facilitators who were drawn from maintenance technical operations (not from the maintenance training department). In addition, Continental used external professional trainers to collaborate with internal facilitators in team-teaching their MRM course. The internal facilitators provided relevance, credibility, and the external trainers provided a professional touch and polished appearance to the course. The combination produced good results.

Other airlines developing new MRM programs use their maintenance training departments to deliver the courses. Yet others use a combination of AMTs and maintenance foremen to team-teach their MRM courses. This team-teaching is based on substantial facilitator training and months' long assignments to the program. The result is high-quality training for AMTs.

In contrast to the internally developed programs, some of the new MRM courses are created, managed, and delivered by external aviation training consultants and trainers. These external courses can accommodate participants from many client organizations at once, or they can be conducted on customers' premises, exclusively for their maintenance employees. Such third-party MRM training may be ideal for smaller companies with few maintenance employees and limited training resources.

Another variation on training delivery is the application of CBI, computer-based instruction technology for team concepts. Already developed and tested [Ref. 2], CBI software can be used to train maintenance employees about team concepts (not team-building techniques themselves), at their own pace and on their own PC computers. CBI can also be used as tune-up refresher training as required.

The New Age of MRM Programs

This newfound interest in open communication for safety improvement is due in no small measure to the generous sharing of information by the

MRM pioneers: USAirways (see Chapter 9), Continental Airlines (Chapter 10), Transport Canada, and The Boeing Company.

As noted previously, the application of open communication training is taking many forms, but several themes dominate North American practice. First, airlines seek programs that are designed for technicians and that contain maintenance-specific cases and illustrations. Second, large airlines prefer to conduct their training themselves, and, where possible, they create their own videos and maintenance case materials. Third, in a pendulum swing away from the Continental model, many airlines train their technicians first.

To Focus on AMTs Alone Is a Mistake

In their efforts to deliver MRM training to technicians, a number of airlines seem to be ignoring their supervisory and management personnel and are not fully involving them in the program. This neglect of foremen, managers, and directors appears to be a policy oversight rather than a conscious avoidance of MRM support training for management. Indeed, there are plenty of places where AMTs and management both participate in MRM programs, but many programs focus (at least initially) on AMTs because months (perhaps even years) can pass before all the company's AMTs can be trained. As a practical matter, this can mean that management training in MRM is delayed for too long or virtually ignored. Recent evidence suggests that leaving management out of the program may have costs that cancel whatever early benefits are achieved.

MRM Training Does Not Make an MRM Program

Chapter 9, defines MRM as reducing errors through better communication and collaboration. This definition is based on the logic that AMTs want to work safely and would collaborate with others to do so, but that they may lack sufficient administrative skills, sufficient trust in others, sufficient communication skills, and sufficient opportunity for collaboration. This logic accepts that all these elements are necessary, but none alone is sufficient—all must be in place for sustained success.

MRM training for AMTs surely improves their communication skills, but it does not (by itself) provide the discipline to administer a maintenance safety program or to provide management support and cooperative resources to collaborate for safety. Even more importantly, MRM training for AMTs does not guarantee sustained trust. Using widespread training as an end in itself is jokingly referred to as "spray and pray." Indeed, if expectations are raised by the training and then not fulfilled, an emotional backlash may occur. To avoid this, the planners and decision makers should embed MRM training in a larger program of open communication for safety improvement, a program that addresses setting and attaining safety goals and that actively and regularly involves maintenance managers as well as AMTs.

Training AMTs Has Immediate Results

The availability of HPIM training materials has fostered interest in providing MRM-type training for AMTs throughout the industry. Several air transport companies have begun the process. Committed to evaluating their progress and their results, these companies have seen clear evidence that attitudes about communication and teamwork improve immediately after the training. Participants' attitudes are also positively related to safety performance [Ref. 3]. Whether this honeymoon effect is sustained depends on many factors, but active management support is paramount among them.

MRM must be understood as a long-term program, taking many years to be fully effective. Undertaking MRM reminds us of the story about dancing with bears: "Once you start to dance with a bear you don't stop when you'd like to, but only when the bear is ready." If your company wants to dance with MRM, be sure all parties are committed to *actively* supporting the program until it is embedded in the company's culture, supported by the whole system as a matter of course. MRM cannot be truly successful as just another spray-and-pray training intervention.

Recent experience with developing a culture focused on open communication and safety is guiding the aviation industry toward involving AMTs early and widely. Several airline companies have embarked on this strategy.

A Case Study in Bringing MRM to AMTs

One company has begun to train its AMTs to appreciate the impact of open communication on safety. Although that company is only beginning its quest for greater safety through better communication, its early results are instructive. The company is addressing its new communication training exclusively to AMTs in the first stages of a larger MRM program. The full story will emerge only after several more years. At the time of the results reported below, the company had not yet taken any visible steps to further promote or encourage the use of the skills and awareness beyond what was covered in the MRM training.

In this case, the company has trained several thousand AMTs over 1-1/2 years. The mechanic's union and the company's management cooperated to initiate the training. Training materials were adapted from Transport Canada's HPIM package, and the company standardized them for its own use, including the use of local case illustrations. Training continued at the local level with facilitators coming from the ranks of both AMTs and their first-line managers. This group of facilitators represents excellent use of local operations experience and leadership abilities. The training was coordinated and supported by the company's training and education department. During the first six months, facilitators collected limited course evaluation data themselves, but over the next twelve months they collaborated with outside evaluators and made extensive use of surveys [Ref. 4]. The results of the first year are reported here.

The core questionnaire used was the MRM/TOQ, the same measure developed 1991–1994 [Ref. 5] and described in Chapter 10. This survey instrument measures attitudes, opinions, intentions, and behaviors expected to be changed by MRM training. Table 11.2 summarizes the topics measured in the current version of the MRM/TOQ. The survey measures all the topics described in Chapter 10, plus a new set of topics measuring respondents' opinions of the company's safety climate.

MRM/TOQ was used to survey the course participants immediately before and after the training (called pretraining and posttraining surveys), as well as at two and six months after their training.

During the first year of training, over 3500 participants attended the training. The vast majority of these were AMTs. The ratio of AMTs (including lead mechanics and inspectors) to all others attending was 7 to 1. By comparison, foremen, line supervisors, and maintenance management accounted for less than 1% of the total.

TABLE 11.2 TOPICS MEASURED IN THE MRM/TOQ SURVEY

Evaluations of the training Attitudes: (I value . . .	Opinions: (I believe . . .
communication & coordination	the safety practices here are available, etc.
stress management	goals are shared, etc.
sharing command responsibility	*Intentions:* (I plan to use this training to . . .)
assertiveness	*Behaviors:* (I have used the training to . . .)

Source: Taylor [Ref. 4]

Additionally, the sample of trainees had the following characteristics. Line maintenance sites account for slightly over half the sample reported below, with the rest coming from base maintenance, tools and materials, and quality control. Day-shift employees represent 40% of those trained. Afternoon and night shifts, plus those on rotating shifts, account for the rest. This is not a young or inexperienced group of AMTs. Their median age is 37 years, and their average experience on the present job is nearly nine years.

Initial Training Evaluations Very High

MRM Training Can Improve Safety and Teamwork

Immediately after the training, 90% of the participants felt that the program had the potential to improve safety and teamwork. Compared with the smaller sample of 300 AMTs described in Chapter 10, this enthusiasm for the course is, if anything, even higher. This is impressive, given the much larger scope of the training program. Figure 11.1 shows the relative enthusiasm for the training as respondents got further away from it over six months. The 90% combined percentage of respondents who say they "agree" (30%) and "strongly agree" (60%) in the immediate posttraining survey is extremely high. Surveyed two months after their training, however, respondents report a combined percentage of only 60%. Six months after their training, the combined percentage of those who agree has dropped to 50%. Why should this drop occur over time? Let's call it insufficient cultural support.

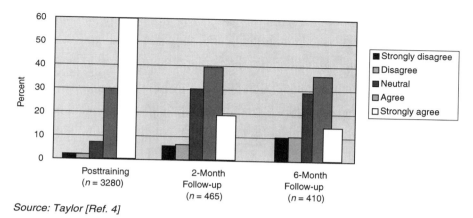

Fig. 11.1 AMTs' opinions: "Training can increase safety and teamwork."

This Training Will Change My Behavior

When the issue of enthusiasm was stated in personal terms (*my* behavior), respondents were overwhelmingly positive immediately following the training even if some hedged a little on their interpretation of substantial change. Figure 11.2 shows the results for the question exploring personal change. The combined percentages (immediately after training) for "moderate" and "large" change categories is over 60%. Thus, a clear majority believed that the training would affect their behavior. Two months after the training, that combined percentage dropped to 46%, and by six months it was down to 42%. What does this one-third drop in personal enthusiasm mean? Respondents appear discouraged with progress and follow-through. Again, what is needed is strong cultural support for change. It's not really personal development that's at stake, it's *organizational* development.

Improved Attitudes

Figure 11.3 graphs mean scores for six attitude and opinion scales measured over time. Comparing AMTs' attitudes immediately after their training with their pretraining attitudes showed significant improvement on three scales.

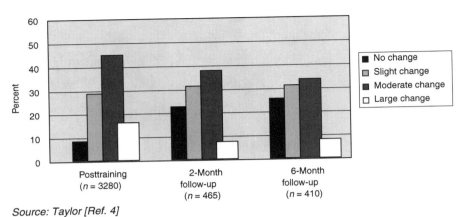

Source: Taylor [Ref. 4]

Fig. 11.2 AMTs' opinions: "Training will change my behavior."

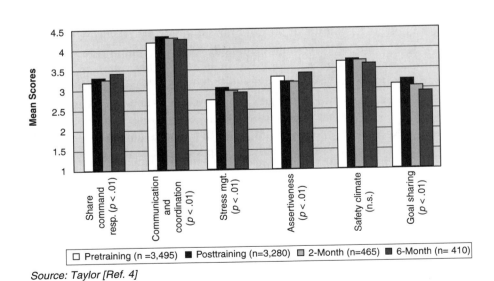

Source: Taylor [Ref. 4]

Fig. 11.3 AMTs' views about MRM topics over time.

Improved Participation, Teamwork, and Stress Management

The statistically significant improvements seen in Fig. 11.3 involve attitudes about "willingness to share command responsibility," "usefulness of communication and coordination," and "recognition that stressors affect decision making" immediately following the training sessions. Further examination of Fig. 11.3 shows that those same three attitudes two and six months later reveal that favorable values remained stable for months after the training. The posttraining increase in these positive attitudes proves that the topics emphasized in the MRM training changed minds in the intended way. Participants learned enough about working with others and about the hazards of unmanaged stress to convince them of the value of these ideas. The stability of these newfound values is evidence that the change sticks— it's not merely a honeymoon effect.

The fourth attitude scale measured "assertiveness" and shows a significant *negative* change immediately following training. That value of assertiveness remains lower but stable for two months after the training; it then increases to a significantly higher level six months later. This pattern differs from data obtained elsewhere, where the value of assertiveness typically remains unchanged immediately after the training. However, the increase in respondents' feelings about assertiveness six months after the training does match those other results. Experience elsewhere shows that attitudes toward assertiveness of both AMTs and maintenance managers do not increase immediately following MRM training, but those positive attitudes do increase in the months afterward [Chapter 10, Fig. 10.5; Ref. 6]. In the present case, MRM training does not emphasize assertiveness heavily nor does it provide skill training to enhance its use. However, simply introducing the notion of speaking up with others, and in turn having them hear you, is an idea that rises in popularity (for the present case) in the months following the training. The spike of improvement after six months reveals an energy to act. That energy might be arising out of frustration over the difference between the heightened desirability of assertiveness and the existing system's tendency to dampen its actual practice.

Figure 11.3 also presents the AMTs' opinions about the company's safety climate and the degree of sharing goal setting and attainment, compared over those four time periods. Let's look at the "safety climate" results first.

Opinion About Maintenance Department "Safety Climate" Unchanged

This opinion scale measures respondents' assessment of the availability and effectiveness of local safety-related practices and policies. Figure 11.3 shows a fairly high assessment of the safety climate overall, but it also shows no statistically significant (n.s.) change in that assessment over time. It is not surprising that ongoing policies and practices are not seen to change during the two days of the MRM workshop, even when safety awareness is a central focus of the training. It is disappointing, however, that the safety climate is not seen to improve in the months after the training. These results plainly show that MRM training, in itself, is not enough to affect fundamental departmental practices. In the eyes of the AMTs, the safety climate did not improve.

Positive Opinions About "Goal Sharing" Decline

Figure 11.3 also shows the results from the "goal sharing" scale. Positive opinions about goal sharing in maintenance increased between pre- and posttraining surveys. Because participants came to the training from different work units and sometimes different shifts, these AMTs had a chance to discuss with one another issues of communication and coordination, as well as lack of awareness, complacency, and other topics in the Dirty Dozen. They could (and did) recognize their common ground, and their posttraining "goal sharing" scores confirm this. Goal sharing has been used in other surveys of AMTs, as reported in Chapter 10. Compared with those earlier surveys, where positive goal sharing opinions *decreased* significantly, AMTs' posttraining scores in the present case reveal a more positive view as a consequence of the training. However, in the following months that positive opinion diminishes, and the six-month score is significantly lower than the immediate posttraining survey. What caused this effect? Goal sharing has not yet become a norm—an ingrained habit—of everyday working life. It has not yet replaced the old culture of management goal tending.

A Recap of Case Results

First, let's review feelings about the course itself. For the some 3500 AMTs reporting their experience with MRM training, there were high initial hopes for its impact, but they fall off dramatically (Figs. 11.1, 11.2). Next, values espoused in MRM training about participation, teamwork, and cooperation—as well as the values of managing stress—all rose significantly after training. Furthermore, these changed values were not a flash in the pan, they persisted over time. Additionally, participants' assessments of goal sharing at work increased following the training, then fell back later. The optimism and enthusiasm immediately after the training was great, but it did not last. Some important values earned higher marks, and stayed that way, but this may not be enough to sustain change. The value of assertiveness first dips, then increases in the months following training. How might this affect changes after training? It could go either way, toward improvement or toward retraction for the maintenance culture as a whole. What management does next will make all the difference.

Performance Changes Related to MRM Training

In the 18 months following the onset of MRM training, both occupational injury and aircraft ground-damage performance improved in line stations. For base maintenance (where the MRM program had less than one year's experience), the results are suggestive but not conclusive [Ref. 4].

Stress Management Improves Safety Performance

Positive attitudes toward stress management two months after training showed the strongest correlations with low rates of injury and aircraft damage [Ref. 4]. Stress management is a topic that MRM training emphasizes; respondents' attitudes show that it "takes."

Stress management is an activity that maintenance personnel can do by themselves, and it does not require the involvement of others (although cooperation may benefit all parties in this regard). The training helps AMTs and their leads improve their individual approaches to handling stress. And this improves safety. But this continued emphasis on working alone may be placing AMTs in the position of not knowing whether or how much the MRM program is working, or whether other people value the lessons of the training as they did. This uncertainty may lead to frustration.

Reported Changes from MRM Training

One question in the immediate posttraining survey and in the two- and six-month follow-up surveys asked respondents to list the personal changes they intended to make following the MRM training. A further question in the two- and six-month surveys asked respondents to list what changes they actually did make as a result of the training.

Intentions to Change

Figure 11.4 summarizes respondents' answers to the question, "How will you use the training on the job?" Results in Fig. 11.4 represent the most important and most frequently stated answers. Although many other answers were given, they accounted for smaller proportions of the total [Ref. 4] and are not included here. For this reason, the total percentage for any of the three surveys does not equal 100%. For the immediate posttraining questionnaire, however, Fig. 11.4 shows that three answers account for almost half of the respondents. "More interaction" (intending to work more closely and cooperatively with others), "Fight complacency" (intending to work more carefully), and being "More aware of themselves and others" totaled 45% of the written answers in the end-of-training survey. For the two- and six-month follow-up surveys, the total proportion drops to a little over one-third (35% and 34%, respectively).

Intentions to Change, Changed

More important than the drop in those positive intentions is the increase in more critical issues shown in Fig. 11.4. The percentage of respondents who said they didn't intend to change, or who made a negative comment about the program or its effects on their future behavior, increased dramatically over time. Those two critical responses together account for less than 5% of the immediate posttraining responses, but they increase to totals of 19% and 27% in the two- and six-month surveys. This is a four- to fivefold increase in negative outlook with the passage of time. Like New Year's resolutions, good intentions definitely faded. Looking behind the summaries at what respondents actually said, many of the negative comments revealed that respondents had tried to change but they were ignored, or not supported, or they actually had been punished when they tried to speak up and become more active. The

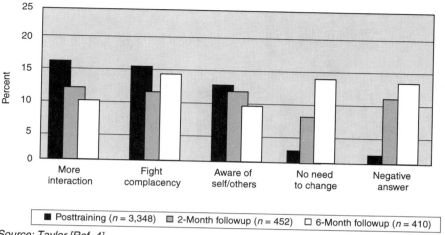

Fig. 11.4 How will you use the training on the job?

culprit is the old culture, exerting its powers of self-preservation, as all cultures do when pressured to change. Cultural change does not come without resistance, ever—not even when everybody seemingly agrees to the change.

Changed Behaviors

Not surprisingly, Fig. 11.5 shows that in the months following training results didn't always match intentions. Although one-sixth of the respondents said their posttraining intentions were to increase interaction and communication with others (Fig. 11.4), Fig. 11.5 shows that less than half that number report actually doing it. On the other hand, Fig. 11.5 also shows that reports of working more carefully (fighting complacency) and being more aware of self and others do more closely match earlier expectations. These results suggest that early intentions to behave differently with others in the workplace may be overly optimistic or naive. Many respondents actually favored only new behaviors that they could adopt passively or by themselves. Stress management is an example of this greater attention to individual and private action. Although too few AMT respondents stated they would subsequently apply the lessons learned about managing stress, the results in Fig. 11.4 show that many did report acting more carefully and self-consciously in the months following training (Fig. 11.5).

Airline Maintenance Resource Management: Improving Communication

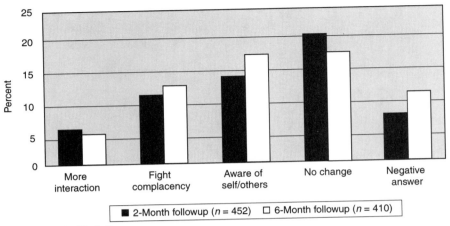

Source: Taylor [Ref. 4]

Fig. 11.5 How have you used the training?

Being thorough, fighting complacency, being aware of one's own impulses and feelings, and observing those of their colleagues—these are useful behaviors that AMTs can do by themselves. But speaking up or initiating work-related conversation with others is more difficult to do without having other, larger changes occurring in the workplace.

Figure 11.5 also shows that reports of "no change" or negative comments about changing are quite high. In fact, the combined percentage of "no change" and negative answers approaches 30% of the total for the two- and six-month surveys, and the proportion of negative answers to the more neutral "no change" increases by nearly one-half between the two- and six-month surveys.

A Parting Thought: The Importance of Implementation

This case of large-scale AMT training in MRM shows how timely and appropriate such a course can be. On exiting the course, AMT participants were enthusiastic about the course and its content, and they were optimistic about its effects on their work environment. This optimism is well placed, because examination of performance results showed that occupational injuries and ground damage to aircraft declined in the 18 months after the MRM training began. Furthermore, those declines were shown to be directly related to improvement in attitudes toward managing stress, a topic emphasized during the training. Without rapid follow-through on the course and its lessons, however, the AMTs experienced a fall-off in enthusiasm and began to show a negative backlash. In fact, the only positive indicator of those measures that shows improvement six months after training is an increased value for "assertiveness," including a willingness to disagree with others and to question others' ways of working. Frustrated with slow progress in achieving the promise of MRM training, a sizable number of AMTs saw a greater need to speak up—perhaps even in anger or frustration—as the only path for improvement. In itself this could be seen as energy for change in the right direction. It could become a lever for positive change in the maintenance work culture. Employee assertiveness should be reinforced and "captured" by management as an important return on their investment in MRM training.

MRM training that becomes an exercise in mere "spray and pray," whether by intention or by accident, sows the seeds of its own discontent. The ideas and behaviors are too liberating to expect participants to watch them erode without reaction.

AMTs are pushed to the front of what is essentially a cultural change, then they wonder where everybody else is—no wonder they get frustrated and discouraged when they don't see support forthcoming. Our advice: Take on the whole system, wholeheartedly. Implement what the training espouses. The big payoff is a culture that breeds continuous improvement in human effectiveness and airplane safety.

References

[1] Gordon Dupont, Special Program Coordinator for Transport Canada, developed the HPIM workshop as described in G.W. Dupont, "The 'test flight' of human performance in maintenance part 2," *Ground Effects* (Vol. 1, Issue 2, 1996).

[2] Kraus, D.C., and A.K. Gramopadhye. "Role of computers in team training: The aircraft maintenance environment example." *Proceedings of 11th Meeting on Human Factors in Aviation Maintenance*. Washington, D.C.: FAA, Office of Aviation Medicine, 1997.

[3] Taylor, J.C., M.M. Robertson, and S.W. Choi. "Empirical Results of CRM Training for Aviation Maintenance Technicians." *Proceedings of the Ninth International Symposium on Aviation Psychology*. Columbus, Ohio, The Ohio State University, 1997.

[4] Taylor, J.C. "Evaluating the effects of maintenance resource management (MRM) interventions in airline safety." 1997 Report of Research Conducted Under FAA Grant #96-G-003, to FAA Office of Aviation Medicine, Washington, D.C., Institute for Safety and Systems Management, University of Southern California, Los Angeles, December 1997.

[5] Taylor, J.C. (in Press). "Reliability and validity of the 'maintenance resources management/technical operations questionnaire (MRM/TOQ).'" *International Journal of Industrial Ergonomics* [Special Issue on Human Factors in Aviation Maintenance].

[6] Taylor, J.C., and M.M. Robertson. "The effects of crew resource management (CRM) training in airline maintenance: Results following three years' experience." Washington, D.C.: National Aeronautics and Space Administration, 1995.

Chapter 12

Recommendation Number One: Change Your Mind

OVERVIEW: The Real Trouble with Maintenance Is Called Culture Lag

The deadly communication problems we've reviewed here are the products of a work culture that has outlived the technology that produced it. What is needed is an evolutionary cultural shift, beginning with a small yet profound change in management thinking, and then continuing with a few concrete actions.

Step one is taking a hard look at the facts.

The Facts Have Changed

Today's reality is different from yesterday's. Maintenance operates in a vastly different world than it once did, a world that is challenging the industry to change the way it thinks about the organization and management of work. The old mind-set is not up to the new reality.

Charles Taylor and the Wright brothers didn't have to deal with the conditions facing the aviation maintenance business today. What's new? Everything: rising numbers of narrow-gauge specialists who speak different technical languages but whose work must somehow be coordinated to ensure the safety of the aircraft as a whole system; the rapid introduction of complex new electronic systems; the sudden increase in new problems with older airplanes; the widening generation gap as large numbers of younger technicians enter the field—less broadly trained, less experienced in aircraft work, less comfortable with command-and-control management practices, and less identified with their elder colleagues, with their employer, or even with their own careers in aviation maintenance. All that, plus a growing factory-hand mentality and loss of professionalism in the ranks. Add the increasing use of outside vendors. Add mergers and acquisitions. Add changes in top management, from airplane people to money people.

These are just some of the footprints marking the increasing pace of change in the industry's entire operating environment, in its technology, and in our society at large. It's not your father's airline any more. Not anywhere. Maintenance has tried to adapt. But it has not succeeded very well.

Old habits of mind and action stand in the way. The culture resists fundamental change. It's cautious, conservative. By its very nature, it's slow to apply lessons learned in other industries, even in other airlines. The maintenance culture rests on a proud tradition that has now become its own worst enemy.

But culture is a product of the human mind. To change the culture, all you have to do is change your mind.

Challenge Basic Assumptions

One ideal has defined and sustained the maintenance culture since its inception, serving as its central organizing principle—that of the self-sufficient, rugged individual. This idea has made maintenance what it is. But what led to so much success for so long is now responsible for the industry's failure to adapt as well as it might. The world has made the lone gun obsolete. But maintenance still acts as if it had not.

The industry's guiding image, the primacy of the isolated part, unfortunately applies a simple cause-and-effect model of the mechanical world to today's work environment and to its people. When there is too much focus on the isolated part, or the isolated individual, then the importance of their interrelatedness is lost. Relatedness must be brought back into view, to restore balance and to allow whole-systems thinking to augment linear thought. This shift in thinking would amount to reorientating the culture around a *new* central idea, the idea of the dynamic whole.

It's not such a new idea anyway. You've heard it a hundred times: The whole is greater than the sum of its parts. Applied to the workplace, this reminds us that the entire environmental-technical-social system is a complex whole, involving the interaction of all of its parts.

Most people readily acknowledge that this whole-system picture is an accurate and valuable way of describing the way things really are. But intellectual acceptance is just the beginning. Maintenance has not yet internalized the full meaning of this wide-angle systems view. The machine model still prevails: one part, one person, one time. The consequences of this kind of thinking are shown throughout this book. Remember the Eastern case?

When Eastern Airlines suffered its infamous O-ring accident, other airlines breathed a great sigh of relief that they did not have this *particular* problem. Virtually no one gave much thought to the larger issues involved, the system issues. But the pattern of poor workplace communication that resulted in the O-ring problem at Eastern is the same pattern that generates other problems at the other airlines. They fixed their O-ring problem, but the problem-making system went unrecognized and unaddressed virtually everywhere. Narrow-focus technical thinking blinded the industry to the larger lessons that the O-ring incident could have taught every maintenance department in the business.

Maintenance culture takes on one symptom at a time quite well, but it is simply not hardwired to handle the larger system issues that continue to produce all kinds of *new* symptoms, endlessly. Recentered around systems thinking, the culture could better prevent mistakes in the first place. Such a culture would include a self-correcting, multimedia communication system.

The one- or two-shot MRM communication training programs also illustrate the case in point. MRM-style programs have been a demonstrable success, by management's own measures. But the MRM approach is not entrenched anywhere as an ongoing feature of a new maintenance culture, a new way of doing business every day. Everybody acknowledges that the good effects taper off over time. But nowhere has management made MRM a continuing part of employee training, much less a part of everyday practice on the job. It's another case of the isolated-part mentality. A brief communication training event does not do much for the workplace communication system. The communication system governs everything maintenance does. A little tweaking does not address the needs for a new system altogether.

Companies like USAirways (Chapter 9) are beginning to strike closer to home, undertaking an array of high-participation programs to improve their communication system: One group tackles the log book, another the general maintenance manual, and so on. Management will have to resist the temptation to stop when each of these initiatives is done; they must instead visibly and consciously tie them together. The mistaken idea of treating individual initiatives as complete in themselves is an example of habitual mechanistic thinking. It's just another take on the norm that says, "Sign it off and it's done!" In a true systems approach to workplace communication, the job is never really done; system-improving activities go on continuously, a central feature of a self-renewing culture. Maintenance will have to make a leap of imagination to get there.

Management must work especially hard to shrug off its long-time enchantment with the traditional, each-thing-in-isolation model. It must reach for a larger point of view that keeps the best of the old but puts it firmly within the context of whole-system thinking. In particular, there is great danger in moving toward lower professionalism and more factory-style management. Two very different choices face maintenance managers today, as shown in Tables 12.1 through 12.3. We hope that the systems model will be the reader's obvious choice.

The specific programs that would be generated to support the new systems model are strongly implied but not minutely described in these pages. We don't want to see you take others' programs and plug them in like so many new parts. The task of cultural transformation requires that the means are compatible with the desired ends. That means *participative* selection and definition of new, *company-specific* philosophies, objectives, work structures, policies, programs, and practices. It has to be taken on as a whole system, *by* the whole system.

Spend a few minutes feeling your way into the very different ideas that are suggested in Tables 12.1 through 12.3. We won't belabor these points here, except to say that management and mechanics alike should get some exposure, some training, and some "safe" hands-on experience with the systems culture—immediately followed with an opportunity to build such a whole-systems culture themselves.

TABLE 12.1 THE NEW SYSTEMS CULTURE

It Is Not	It Is
1. Simply problem-centered	1. Purpose/product centered
2. A canned solution invented elsewhere	2. A continuous search for your own solutions
3. An add-on program	3. Whole-system change
4. Oriented toward certainty, finality, stability	4. Tentative, flexible, dynamic
5. The same old stuff, dressed up	5. A new way of working life

Source: Adapted from Christensen and Taylor [Ref. 1], p. 3

TABLE 12.2 THE NEW SYSTEMS MODEL

It Is Not Mechanistic	It Is Systemic/Organic
1. CONTROL: Of people, largely from above	1. CONTROL: Of safety, largely from within
2. COMMITMENT: Directed toward boss and task	2. COMMITMENT: Directed toward customer and aircraft safety
3. COMMUNITY: Us *vs.* them Adversaries Closed system	3. COMMUNITY: Us *and* them Partners Open system
4. CONSCIOUSNESS: Low psychological investment Individual in isolation Parent-child Fear, distrust	4. CONSCIOUSNESS: High commitment Individual in relation Person-person Courage, trust

Source: Adapted from Christensen and Taylor [Ref. 1] p. 5

TABLE 12.3 TWO CHOICES FOR MAINTENANCE MANAGEMENT

Mechanistic Practices	Systemic Practices
1. Treating people as extensions of the airplanes, as tools	1. Treating people as complementary to mechanical things, as the masters
2. Seeking to optimize technology	2. Seeking joint optimization: social *and* technical systems
3. Maximum task breakdown, narrow specialization working in isolation	3. Optimum task grouping; multiple, broad skills, working in teams
4. People as expendable, easily replaceable spare parts	4. People as key resources with further potential
5. External controls: supervisors, outside experts, procedures	5. Internal controls, self-regulating subsystems
6. Tall organization chart, autocratic style	6. Flattened organization chart, participative style
7. Organization's purposes only	7. Members' and society's purposes too
8. Discouraging innovation, initiative	8. Encouraging innovation, careful experimentation
9. Competition, gamesmanship	9. Collaboration, teamwork

Source: Adapted from Eric Trist [Ref. 2]

Many consultants offer orientation programs in this way of working, and some new-culture companies encourage outsiders to come in for a good look. Some excellent books and articles on the subject are listed in the references at the end of every chapter.

Tables 12.1 through 12.3 are all expressions of the same thing, the choice of cultural values facing today's aviation maintenance managers. What has been described in the preceding eleven chapters should lead you away from the mechanistic culture to embracing a systems culture that is better suited for today's world—the world of complex interrelatedness wherein every "part" is known to have a powerful (if sometimes hidden) impact on every other part.

The maintenance culture was once built around the ideal of the all-purpose, generalist mechanic. It worked. It made maintenance the proud profession that it has been for a long time. But the old culture is already slipping the *wrong* way—toward a kind of factory system, reducing workers to mere specialized parts that have to be controlled from outside and above. This is not progress, but its opposite. It spells trouble in more ways than you can count. Don't go down that road. It's a dead end.

The culture we advocate restores the old ideal of the self-sufficient workforce, but redefines workforce to mean the self-sufficient team, the self-managing work system. It is a mistake to fragment the workforce into an unrelated collection of isolated specialists. This mistake can still be corrected; indeed, it *must* be corrected before things get too far out of hand. Put your specialists together, build multifunction teams, and tackle the complexities of the technical system with a social system designed to match.

Stop Working at Cross-Purposes

The standard line is "speed and accuracy" or sometimes "accuracy and speed," as if these two goals were equal and never in conflict. When you can't have it both ways, which goal comes first? AMTs usually see their core mission as safety (accuracy)—assuring the airworthiness of the planes they return to service. But management often appears to favor speed—putting the planes back on line on time. Real airworthiness is hard to measure—unless there's an accident. The next-best measure of safety is conformance to established standards and procedures, a measure that is almost as difficult to track, since every plane requires a different fix and every fix involves its own complex set of standards and procedures. Not even the most meticulous paper trail can say that each and every standard was actually *met,* only that it was addressed.

Nor can any after-the-fact quality control inspection put quality *in* or even ensure that the mechanic followed the book in every instance. In one case discussed earlier, a crack had been plated over (Chapter 6). Quality control couldn't see it and thus couldn't fix it. The bad work could only have been prevented or done over had quality control been there the whole time, looking over the test operator's shoulder, *and* had the inspector himself been alert to the proper standards and procedures to be followed for that kind of repair. Even in the best of present circumstances, it's hard for mechanics and inspectors to see themselves as a team, dedicated to the same purpose.

Quite the contrary, setting them up as two different functions isolates mechanics and inspectors from each other, creates a source of conflict, and falls short of the kind of quality assurance that results when qualified mechanics are assigned to the work together, sharing responsibility for the quality of their work. This point is of utmost importance: In maintenance, quality means safety, and safety is the core mission for everyone.

This mission, the very reason maintenance exists, is to make sure planes are safe to fly. But managing for safety is nowhere near as easy as managing for other goals, timely turnaround, for example. One report we covered earlier details the excellent results obtained when inspectors and mechanics worked together with the express goals of meeting *both* goals better: quality *and* schedule (Chapter 4). They set objectives together, they supplied support to each other, they measured results jointly. And they succeeded in making speed without sacrificing safety. When speed or cost gets more attention than safety, the mission gets confused. *People* get confused. They lose their focus.

Maintenance needs a thorough, all-employee discussion of its true mission, a revitalized commitment to the safety imperative, and a better understanding of its practical relationships to time and cost considerations. Safety, speed, and cost are all involved in the mission, of course. But everyone must be clear about what's first, second, and third in decision making, and why and how they are expected to maintain a proper balance.

Ensuring mission clarity is a subtle task. The mission should define the inseparable linkage among safety, time, and cost, while keeping safety always in the forefront.

Perhaps the mission should be stated and understood as "keeping the planes in the air." Making sure that planes don't crash must obviously take first place in everybody's mind. Keeping planes in the air by minimizing downtime must be understood as a laudable, but secondary, consideration. Managing expenses also helps ensure that the planes stay in the air—it helps keep ticket prices down, thereby filling the seats and keeping the airline in the sky. These factors are all involved, and they must all be included to make the mission both meaningful and practical.

Furthermore, managers and employees alike must come to see how productivity (speed and cost) and quality (safety) are often *not* contradictory goals, but direct contributors to each other's success. A truly quality job is also the most time- and cost-efficient job; the most truly productive work is quality work. Shortcuts, even safe ones, often waste more time and money than what appears to have been saved by not doing the whole job right the first time. This is old stuff; everybody knows it. But not everybody acts as if they know it.

In one industry after another, it has long been shown that a shared commitment to quality first *always* generates productivity. It doesn't often work the other way around. What's more, "safety first" is a mission that gets the wholehearted support of mechanics, managers, customers, and regulators alike. It's a mission that draws all parties into the fold, it has sales appeal, and it buys invaluable support and commitment from inside and outside the maintenance enterprise.

With its mission properly understood, maintenance can follow through with a properly supportive operating philosophy, a clearly stated set of decision guidelines. And it can back that up with a compatible list of specific and measurable key objectives. A culture in which any of these parts is not working in clear harmony with the rest is a culture in trouble, trouble that will be felt throughout the whole system. Maintenance errors would not be the half of it.

If maintenance can evolve a healthier culture for itself, it will be doing a very good deed for the larger culture around it—for the whole world, in fact. Cultural change is a far more noble pursuit than merely making things better for yourself. Make things better for everyone; make things better, period.

Reconceive Your Objectives

A transformed maintenance culture uses participative methods to define and achieve three kinds of objectives: technical, economic, and social system objectives. These objectives lay out the measures by which the whole culture is expected to fulfill its mission.

An organization's objectives shape the culture as much as they express it. The objectives actually pursued, reinforced, and rewarded tell you pretty much all you need to know about the culture and where it's headed.

To talk about safety or human resources in high-falutin terms, then to scrape your knuckles raw just for the sake of speed—this culture is driven by the clock, not by passenger safety. And this culture suffers from an advanced case of hypocrisy.

Maintenance is doing just fine by the clock. It needs to make sure that its safety objectives stay in front of speed and cost. Most of all, it needs to set new objectives for building a social system that is fully able to deliver, now and in the future. A new strategy toward human resources management is needed.

New Strategy Overdue

In its overall management strategy, maintenance has clung to the practice of making incremental technical improvements and, in recent years, of undertaking scattered small-scale social-system experiments in the form of MRM-style communication training, roundtable problem-solving meetings, or participative paperwork design.

Initial results have been very good. Both morale and actual job performance have improved. But maintenance has not yet made sure that its socialsystem improvements will be sustained and improved over time. It has not yet made these new *programs* into a new *culture*. That will take a larger commitment of time, energy, and money. True cultural change does not come easy and it does not come cheap.

MRM's lack of staying power tells the story. Despite its initial successes, MRM hasn't yet taken root. Why? Because it threatens the old culture's existence. MRM is foreign, introduced into a culture that would quite literally have to die for this new kind of thinking to thrive. The resistance is natural. Every culture resists change at its core; its powers of self-preservation had to be strong for it to hold its own in the world. Even when a change of heart has become imperative, when it is seen as desirable and when it's wanted by the culture's members, even then inertia keeps the old culture rolling. It lives in its members' minds, commanding their continued obedience, often given without their conscious consent. It's hard to escape the pull of habit.

At the level of culture, even the most needed, most rational change is hard to bring off. It takes a strong push from outside pressures and the strong pull of an attractive new idea to bring it about. Both kinds of energy are now pushing and pulling at the airline maintenance industry.

The purpose of this book has been to bring both the threat and the promise of cultural change into greater awareness and to encourage maintenance to step up to the need to change before conditions reach the crisis stage. If you wait for the crisis to come to a head, you're likely to adopt either or both of two panic "solutions"—a hunkering down, ever-more stubborn clinging to the old ways, or a blind rush into superficial attempts to copy those who've adapted successfully. Wrong answer, either way.

With time running out, maintenance must proceed with all due speed to confront its own core values and to take on the work of real reform. One-shot training won't get this done. Focusing only on mechanics, and not on the support people or on the maintenance bosses, won't get it done either. The best learning and the most productive mode of working life springs from new ways of organizing and managing the work, based on new ways of thinking about the work.

A culture based on systems thinking makes innovation permanent, institutional. It combines ongoing training with daily whole-crew meetings, team-based work assignments, big-picture briefings, participative goal setting, and high-involvement problem solving. These and other multifaceted, high-communication practices are at the *center* of regular working life, not featured as extracurricular, short-term programs.

To be fit for these times, new organizational and management strategies must incorporate systems thinking into the very fabric of the culture. Work structures and support systems must build professionalism into every job (see Chapter 7 on competence, centrality, commitment, and control). Self-regulating work teams with a new pattern of management support must be created, providing the entire workforce with more and better information, constant improvement in technical and social-system knowledge, more power in decisions affecting it, and new rewards for effective teamwork in fulfilling the mission and meeting mission-based key objectives [Ref. 3]. Taken as a whole, these measures define a culture that is far more productive and adaptive to change—a culture that will survive and thrive in an environment of continual and complex change.

The alternative strategy is to *control* individual parts, one at a time, from the top. This might seem tempting. The systems strategy, however, is to control the whole work system, all at once, through a range of high-involvement methods, with appropriate *support* from the top. Jack Sherwood describes management in a new role, featuring the "four Ls": leadership, looking outward, letting go, and lending support [Ref. 4].

In its leadership role, management would shoulder responsibility for keeping the new vision alive, for managing the environment, and for anticipating and managing the future. Leadership includes building a coalition of understanding and support among customers, suppliers, corporate staff, regulators, and other stakeholders whose backing is required for the new culture to succeed. Only political skill and persistence can overcome resistance to the necessary redistribution and sharing of power.

In managing the future, the role of the leadership is twofold. Focusing attention on the future provides the awareness necessary to keep current and remain sensitive to changes in market trends and other environmental changes and demands. It also means keeping attention focused on the new vision of the culture as an attainable future state, especially through the difficult times of getting the new system in place, when midstream troubles sometimes cause the workforce to become dispirited, supporters to fall silent, and critics to gloat.

Looking outward means that managers turn their attention to concerns that lie outside the regular work process rather than attending to the demands of the work itself. The work itself is left to the people who do the work. Instead of asserting direct control over work tasks and work assignments, management makes sure that three things happen at the boundaries: Inputs meet the needs of the work system; outputs meet the needs of the customer and other stakeholders; and environments are managed in such a way that people get what they need to do their jobs with minimum interference.

In letting go, managers give up the illusion of personal control. Instead, managers work for better *system* control by letting go of three things: control over internal methods and details (which passes to self-regulating workteams); management specialization (in favor of membership in a multifunctional management team, mirroring the multifunctional workteam structure); and

Recommendation Number One: Change Your Mind

status symbols (such as separate dining facilities or reserved parking). This, of course, assumes that the philosophy behind a company's labor relations policy and the union's contract *both* encourage the workforce to take control of internal methods and to collaborate closely with staff specialists and management. Unions, too, will have to let go of certain old ideas, as they have in many industries [Ref. 5], in the best interests of their current members.

Lending support, according to Sherwood, means shifting management's role to providing service and support functions to its self-regulating work teams. When managers *do* step inside the work system, they serve as teachers or facilitators to the work of others—encouraging learning, suggesting new ways of problem solving (including conflict resolution), and stimulating creativity [Ref. 4].

None of these ideas is unheard of in maintenance. The challenge is to bring them into the center of organizational life, to make them the heart and soul of a new workplace culture. Appendix A outlines eleven principles for creating a high-communication, high-performance culture. Appendix B supplies a list of supportive programs, tactics, and communication media that might arise with such a new culture.

Appendix B names only the new-culture innovations that have already been discussed in earlier chapters. There's no question that maintenance has the ability to do these things—most of them are already being done somewhere in maintenance. The question is whether maintenance can turn these experiments into the basis of a new culture. The question behind the question is whether maintenance can resist the inclination to choose the self-limiting mechanistic culture and thereby avoid the real work of significant and continuing cultural transformation.

There's no trouble with workplace communication in maintenance that a new cultural orientation can't handle. Where to start? Start by changing your mind.

Achieving an Effective MRM Culture

Informed by a new central organizing principle, "relatedness" or systems thinking, maintenance can design and implement an ongoing multimedia communication program that is based on cross-functional, face-to-face meetings supported with print, intranet, and e-mail, along with new audio-visual aids, training, job design, policy and leadership innovations.

It doesn't matter which comes first—the new framework for thinking or the new programs. So long as both perspectives are kept in mind, principle and program will work together to create the new culture demanded by the times. Indeed, the new approach to workplace communication embodies the new culture. It *is* the new culture.

If these ideas seem to entail lot of overlap and redundancy, that's on purpose. The reason is pretty much the same as for the redundant systems that are built into airplanes: safety, our insurance that the system as a whole will perform as promised. A well-engineered and well-managed system of communication, moreover, does not wear out as airplanes do. A good communication system is self-maintaining and self-improving, continuously. It readily adapts to new conditions, even to the unexpected. And it amounts to the best guarantee you'll ever get that the work itself will be done the right way, every day. What more could you ask?

Here in a nutshell is the new program we recommend, giving an overall sense of order and direction to the list of initiatives shown in Appendix B:

New Work System Understanding

Maintenance should thoroughly investigate the socio-technical systems approach to the design and management of a multifunctional, team-based work system [Ref. 1; 6; 7]. Use a qualified outside consultant to provide your orientation, to guide your readiness assessment, to facilitate your decision process—who, when, where, how, and if—and to coach you through the process. We'd like to say you can handle this on your own, but experience proves it just can't be done. Get help to get going, *then* proceed on your own.

All such work systems are custom-designed by the system's members. Multifunctional teams meet daily and work together throughout the day in various configurations, depending on close collaboration to guarantee high performance in fulfilling their shared mission and objectives.

No other approach can compare to the results that a properly designed and progressively managed team system delivers day in and day out, simultaneously delivering the goods in quality, productivity, cost control, learning, innovation, and workplace satisfaction. It's not a panacea, but it's as close to one as you're going to get in this life.

Workplace Meetings

Supported with a variety of media, the new maintenance culture would feature frequent face-to-face meetings of many kinds. Here's our to-do list, starting at the top of the hierarchy but focused on empowering the people at the so-called "bottom"—home of the most important people in any organization, the people who do the work that the organization exists to do.

State of the Company Meetings

Led by the chief executive, at least annually. A series of live presentations is ideal; filmed or videotaped presentations at all locations is an acceptable alternative. Even if the audio-visual route is chosen, a live question-and-answer period would be provided, with one or more on-site executives fielding the questions. The CEO would address overall performance against primary objectives, in the context of fulfilling the company's mission in a changing environment.

After the talk from the top, employees would meet in small groups to discuss how the issues raised might affect them and what they might be able to do about it. This could make up the core agenda for one of their regular crew meetings, discussed below.

Also very important: follow-up. A printed summary of the highlights of the boss's remarks would be distributed to all employees as soon as possible following the State of the Company meeting. The summary might take the form of a free-standing special report, the lead article or only article in a regular newsletter, a bulletin board posting or all of these—the shortest version being a quick-read bulletin directing employees to the more-complete newsletter summary and/or to the comprehensive special report.

All versions of the after-meeting report would include a summary of the questions and answers of wide interest, the maintenance executives' thoughts on the meaning of the CEO's message as it applies to maintenance, and the crew-level discussions and action plans. The whole nine yards.

This same general model would apply to regular meetings of three other kinds. The object is simple: to create and sustain attention, understanding, commitment, and effective action at every level. Everyone is centrally involved in managing the company's success, starting at the level of their own jobs. Leadership merely lights the way.

State of Maintenance Meetings

These would be led by the maintenance vice president at least semiannually. Reporting on the maintenance department's special mission, objectives, and action plans, these meetings would follow the same pattern as for the CEO meetings—live or on-screen, including the linkages between companywide issues, general maintenance issues, and site-specific and crew-level issues.

Questions and answers, small-group discussions, and printed summaries would be included as a matter of course—it's culture now. It's the way these things are done, at every level.

State of the Site Meetings

Led by the top site executive and involving other site managers and union officials, the presentation and/or questions and answers might profitably be held quarterly—again supported as before with small-group discussion and timely follow-up summaries.

Crew Meetings

These may be the most important meetings of all, since it is at the level of the work team or shift crew that the rubber hits the road. Crew meetings would be led by the foremen and lead mechanics, right on the shop floor, with short daily or at least weekly discussions of immediate performance issues in safety, turnaround time, cost management, and the quality of working life—all this in the context of the employees' shared mission and site-specific objectives.

Support at this level would include prepared outlines or talking points for the foremen, on issues that they and their crews have had a part in identifying. Foremen and other team members too would be trained and retrained in conducting effective, high-participation meetings. Perhaps their boss's bosses would follow the same agenda with the foremen first, providing a direct model for sharing the information, dealing with questions, and leading problem-solving discussions. Foremen could be provided timely performance data that they could reference in their meetings with employees.

Workplace Performance Information

The same kind of performance data could be used to update large, poster-sized graphs on the shop floor or in the break rooms. The graphs would continuously measure performance against the crew's objectives. Objectives and measures would be set with the active participation of the mechanics, as part of their regular workplace meetings. Key goals and results could also be summarized in brief paycheck stuffers, with appropriate management comment, thus keeping the right issues upfront and reminding everyone about the inseparability of objectives, jobs, and paychecks.

At workplace meetings, especially, employees and their foremen should be given official permission and the means to pose questions, define concerns, and propose suggestions for higher-level management. Naturally, this feedback loop would include responsive action by management, including (yet once again) committing themselves in print, soon. And not everything should have to go up the hierarchy: Spell out what must go up and what can and must be handled by the teams themselves or by their members individually. The more that can be done at the grassroots, the better. Push important decision making downward, along with the information and training needed to ensure *good* decision making.

Responsibility for Identifying and Managing Errors

The "crackdown" philosophy so visible in the 1950s and 1960s has not gone away (see Chapter 4). It is alive but undercover; no intelligent 1990s manager or consultant would extol a blame-and-punishment approach to maintenance error. But it continues, because management has not been directed to a readily available and satisfactory alternative—until recently.

"Learning from our mistakes" is a central theme in the MRM approach to safety improvement. This theme has been brought into sharp focus by the Boeing Maintenance Error Decision Aid (MEDA), mentioned in Chapter 11, as well as by similar programs intended to "improve the fidelity and increase the frequency of maintenance error investigation" [Ref. 8]. Specifically, such programs are intended to shift the perspective from finding *someone* to blame for an error to understanding the *process*, the system, that leads to specific types of error. Abandoning the crackdown management style requires the mutual expectations of AMTs and their managers that they will work together to control human error by addressing its root causes, not attacking its perpetrators. Such expectations are not realistic unless they are founded in a confidence that people have the skills as well as the desire to collaborate. MRM and maintenance error investigation are both fundamental cornerstones in achieving a systemic, error-reducing culture in aviation maintenance. And no, this would not excuse gross malfeasance.

Training in People-Centered Management and Error-Process Investigation

A first step in replacing the crackdown approach would be the kind of leadership training we described in Chapter 11 being offered by The Boeing Company. In its two-day introductory training for management, Boeing's agenda covers the central role of the AMT, the necessity of effective workplace communication, the requirements of error management (MEDA), and the use of systemic improvement methods.

Roundtable Meetings

"Roundtable" is a term used at USAirways for groups examining and discussing the causes and solutions for maintenance incidents involving human error [Ref. 8; 9]. These meetings have been useful in reducing future errors. The roundtable discussion format represents something larger for our purposes. It's a good example of a single-issue, problem-solving discussion. Single-issue, problem-solving groups should be chartered as needed to respond to special needs, after which they would be disbanded. New groups would be formed to grapple with new problems or to capture new opportunities as they arise. We recommend *against* standing committees but *for* single-focus special teams—even when "solved," problems like log book design always benefit from additional work. In every case, the groups should be formed with new members joining experienced members from earlier roundtables. The groups should always be multifunctional, multilevel, and cross-shift in makeup. They should get training in group problem solving and statistical analysis. And they, too, should issue frequent face-to-face briefings with their fellow employees, complete with questions and answers, discussions, idea collection, and feedback. Printed reports should be prepared to summarize all of this and to direct employees to any new pages in their general manuals, job descriptions, log books, company policy publications, turnover reports, training materials, or other official new documents.

Meeting Training

Again, for emphasis, *everyone* should be trained in meeting skills, including various communication, conflict resolution, and problem-solving techniques. This kind of training should become a regular fact of life for maintenance people at all levels, with the training held at frequent intervals, introducing new and higher-level skills over time, and including short classroom learning followed immediately by hands-on application—on the job, in the work team meetings, and/or in the single-issue roundtable meetings. Early on, the meetings in which the new skills are to be practiced should be facilitated by professional trainers or by savvy part-time facilitators drawn from other departments or management levels. As soon as they're ready, the foremen should be expected to facilitate the application of this new learning in their work crews. These tasks should be a new line in their job descriptions.

Technical Training

Developing employee competence is an objective that should be upgraded to the level of "continuous," and it should include continuous development of technical skills as well as human interaction skills. Besides on-the-job training (OJT), audio-visual (including computer-based training), and classroom training, the various regular meetings are an ideal forum for introducing new knowledge on a continuing basis, in small and easily digestible bites. As with every other objective, the competence-development principles and programs would be well supported by such media as higher management statements, print, and structured group discussions.

Communication Support Staffing

Maintenance should take steps to ensure that the foregoing communication support is on-line and effective. Hire a qualified consultant or permanent communication professional to set up the multimedia support program, to make it a high-quality program, and to teach others how to do more and more of it for themselves. As before, we'd like to be able to tell you that you can pull all of this off very well without the help of a communication specialist, but it's just not realistic to let you think so. Get help; just be sure to resist the temptation to make communication a one-person or one-department job.

A Final Thought

Communication skills should be added to everyone's job description and supported with training, real information, feedback, and encouragement. Unless this high-involvement answer gets structured in, the "communication question" will get louder and louder over time. Organizational inertia will pull the system back to the past—back to its error-generating ways, its habitual pattern of anemic communication, and its predictable increase in airplane accidents. Kick the habit, starting now.

How? Adopt a simple four-point strategy: Get your head on straight, get professional help, get a well-planned communication system institutionalized, and get a healthy handful of workplace communication skills into everybody's repertoire. Maintenance is already taking its first tentative steps in this direction.

Take the next steps, together. You're up against a whole-system challenge; it requires a whole-system response.

References

[1] Christensen, T.D., and J.C. Taylor. *STS Handbook*. South Bend, IN: STS Publishing, 1991.

[2] Trist, Eric. "QWL in the 1980s." In H. Kolodny and H. vanBeinum (eds.), *The Quality of Working Life and the 1980s*. New York: Praeger, 1983.

[3] Lawler, III, E.E. *High-Involvement Management.* San Francisco: Jossey-Bass, 1986.

[4] Sherwood, J.J. "Creating work cultures with competitive advantage." *The Journal for Quality and Participation, Special Supplement* (December 1989): 14–25.

[5] Rosow, J.R., et al. "Achievement at Saturn Corporation (Part II, p. 31)," *Participation, Achievement, Reward: Creating the Infrastructure for Managing Continuous Improvement.* Scarsdale, NY: Work In America Institute, 1997.

[6] Taylor, J.C., and D.F. Felten. *Performance by Design: Sociotechnical Systems in North America.* Englewood Cliffs, NJ: Prentice Hall, 1993.

[7] Drury, C.G. "Chapter 6: Work design." In *Human Factors Guide for Aviation Maintenance.* Washington, D.C.: Department of Transportation, FAA, Office of Aviation Medicine, 1996. [The *Human Factors Guide for Aviation Maintenance* is also available in *Human Factors Issues in Aircraft Maintenance and Inspection '98* (CDROM), Washington, D.C.: FAA, Office of Aviation Medicine (http://www.hfskyway.com).]

[8] Marx, D.A. "Learning from our mistakes: A review of maintenance error investigation and analysis systems." In *Human Factors Issues in Aircraft Maintenance and Inspection '98* (CDROM), Washington, D.C.: FAA, Office of Aviation Medicine (http://www.hfskyway.com).

[9] Kania, J.R. "Panel presentation on airline maintenance human factors." (CDROM) *Proceedings of the 10th Meeting on Human Factors Issues in Aircraft Maintenance and Inspection.* Washington, D.C.: FAA, Office of Aviation Medicine (January 17–18, 1996).

Appendix A

Principles of Cultural Redesign

Albert Cherns [Ref. 1; 2] has provided us a list of eleven proven principles for the design of a new high-communication, high-involvement work culture, based on systems thinking.

Compatibility

"Means should fit ends," Cherns says. "Only participative design could lead to a participative organization."

Cherns says this principle is often misunderstood, but that it is critical to get it right. It means you don't have a "participation department" ("a bureaucratic instrument to de-bureaucratize"), and that you *do* have a tough-minded, representative design team whose members truly represent their functional areas of expertise and their colleagues, who understand "joint optimization," and who live by the same rules of open communication and consensus decision making they hope to instill in the entire organization.

Minimal Critical Specification

"This principle has two aspects, negative and positive," Cherns writes. "The negative simply states that no more should be specified than is absolutely essential; the positive requires that we identify what is essential."

"While it may be necessary to be quite precise about what has to be done," for example, "it is rarely necessary to be precise about how it is done" or by whom.

Variance Control

"Variances should not be exported across unit, departmental, or other organizational boundaries." Designers must identify social and/or technical means to ensure that variances are controlled as close to their source as possible.

Boundary Location

"Boundaries should *not* be drawn so as to impede the sharing of information, knowledge, and learning." Separating inspectors from the mechanics whose work they inspect, or separating planners from the mechanics they plan for, are both examples of organizational nonsense, which leads to endless conflict, errors, and confusion of ownership.

Information Flow

"Information for *action* should be directed *first* to those whose task it is to act," the people who do the work. In other words, AD notes should go to *both* the mechanics and their engineers to jointly and correctly interpret the work to be done.

Whether it's production, quality, safety, planning, or cost control, "it is no use holding an individual or a team responsible for any function and then doling out information about its performance in arrears and through a higher authority." Those who need the information (to regulate *their own* performance) must also be included in the design of their own information systems.

Power and Authority

"Those who need equipment, materials or other resources to carry out their responsibilities should have access to them and authority to command them." Likewise, workers should be empowered with the knowledge and authority to handle their own fire fighting rather than having to rely on top management command.

Multifunctionalism

Organizations adapt more effectively to the challenges in their environment (including other parts of the organization) when they're treated as adaptable multifunctional *organisms* rather than as a collection of single-function *mechanisms*.

Adding experts to handle a given problem is the mechanical response (a specially designed gadget, say) that only adds to line/staff confusions, paperwork and problems of organizational integration. Training is the organic response—a new trick—enlarging the repertoire of individuals and teams without complicating the allocation of responsibilities.

Support-System Congruence

Support systems should be designed in the same way and to advance the same ends as the basic work system. The ideal would be to enable work units to operate as self-contained "mini-companies," complete with their own quality, purchasing, marketing, planning, and financial control systems, with the fewest possible constraints.

Reward systems that "pay for what you know rather than for what you do" are now common in socio-technically designed organizations, particularly in high-tech operations, where operator errors can be calamitous and where it is most easy to see that people are valuable to their organizations in direct proportion to what is in their heads, not just their hands. And group bonus programs like gain-sharing are often used to reward improvements in a way that also reinforces teamwork.

Human Values

Cherns says that an overarching principle in new age organization design is to provide a high quality of working life, building into the work structure such basic human values and desires as choice and room for individual differences, for variety, for responsibility, and for community.

Transitional Organizations

Cherns states that the "transitional organization" embodied in the design team is "both different from and more complex than either the old or new" system. Its membership, its process, and its every action, he says, is vital to diffusing the new values.

Cherns suggests that the design team, rather than "experts," should deal with the stresses of start-up, shutdown, layoff, selection, and training as opportunities to learn and "to demonstrate the reality of the new philosophy."

Incompletion/Continuous Improvement

To design for incompletion is to design an open system—a system that's always ready for change, expecting change and even thriving on change. Incompletion is the key to adaptability, innovation, and continuous improvement.

Incompletion means that the design itself is always considered incomplete and open to change. It means that the work structure design is only an outline and not a detailed manual. It means that the design will only describe the basic structure of the organization and the means for changing that structure. And it means that job design will be left for the job holders themselves to complete. This in itself is the primary secret to creating continuous improvement in quality, productivity, cost effectiveness, teamwork, and everything else.

References

Cherns, A.B. "Principles of sociotechnical design." *Human Relations.* (vol. 29 1976): 783–792.

———. "Principles of sociotechnical design revisited." *Human Relations.* (vol. 40 1987): 153–162.

Appendix B

Some New-Culture Programs Already in Place

Here's a list of new-culture compatible initiatives already in place in maintenance somewhere. Real transformation will include putting many of these ideas, not just a few, into the regular way of working life in airline maintenance. As mere add-on programs, results are limited and short-lived.

1. Team-based organization: multifaceted membership.
2. Goal sharing.
3. MRM-style communication training.
4. Employee surveys on attitudes, error causes, and possible correctives.
5. Participative problem solving, roundtable meetings.
6. Team meetings, including cross-shift members.
7. Overlapping shifts at turnover.
8. Participative redesign of log books, general maintenance manuals, work cards, turnover communication, engineering orders, ADs' technical publications and policy manuals.
9. Rewrite job descriptions to include communication and collaboration expectations.
10. Improved employee selection criteria, adding social skill requirements.
11. Coordination of maintenance control, planning, and records.
12. Redesign transit check procedure.
13. Training in error-free paperwork.
14. Replacement of microfilm with user-friendly, real-time computerized database, participatively designed.
15. Redefinition of mechanic-inspector relationship.
16. Formal training and OJT on unfamiliar airplanes; recurrent training to ensure maintaining industry technical standards.
17. Expand AD and other important message transmission to include multimedia approach: pictures, audio-visual, meetings, demonstrations, etc.
18. Address the AMT/contract maintenance split with "one-system" strategies; include maintenance in vendor contracting process.
19. Disallow paint-over before inspection.
20. Participatively define safety mission and communicate it continuously *vis a vis* time and cost goals.

21. Get senior mechanics into night-shift overhaul and mentoring newer AMTs.
22. Improve skill and knowledge in testing.
23. Restore professionalism through job design, training, communication, and rewards.
24. Provide regular communication of big picture concerns in technology, company business, industry developments, regulatory and societal climate, etc.
25. Demand higher employment standards, and licensing, at outside repair stations.
26. Post charts showing results versus objectives.
27. Widen and strengthen face-to-face, work-related communication.
28. Dramatize top management support for training, communication initiatives.
29. Get AMT involvement before issuing formal engineering orders.
30. Set up an AMT-to-management telephone hotline, with published follow-up or questions or suggestions and management's responses.
31. Reduce the number of maintenance checks (and their paperwork) on through-flights.
32. Seek all-employee understanding, participation in culture change as best response to new concerns about safety, speed, cost, professionalism, and the quality of working life.

Index

A&P Mechanic. *See* Airframe & Powerplant Mechanic
Accidents, 2
   Air Ontario (1989), 95–100
   Aloha Airlines (Maui, 1988), 21–28, 54–55, 59, 125
   Atlantic Southeast Airlines (Carrollton, 1995), 69–71
   communication system as cause, 2, 21
   Continental Express (Eagle Lake, 1991), 21–22, 28–33
   Continental Express near miss (1992), 33–34
   Eastern Airlines near miss (1983), 74–75, 165
   independent repair stations, 67–77
   quality control as cause, 67–69
   ValuJet Airlines (Atlanta, 1995), 67–69, 169
   ValuJet Airlines (Miami, 1996), 71–74
ADs. *See* Airworthiness Directives
Air Canada, participative management, 45
Air Ontario accident (1989), 95–100
Aircraft Maintenance Engineer (AME), 79–80
Aircraft Maintenance Technicians (AMTs)
   characteristics, 12–13
   commitment, 83, 85, 87–88, 167
   communication, 7–9, 11–12, 60, 63, 77, 139
   communication training, 19, 47, 143–161
   competence, 83, 84, 87
   complexities of job, 1–3
   consciousness, 167
   control, 83, 85, 87, 167
   mistakes, reasons for, 15–18
   MRM participation, 113–114, 135, 149, 150–152, 165
   new systems approach, 166–182, 185–187, 189–190
   professionalism, 53–64, 77, 79–88, 125, 173
   selection criteria, 11–12
   sense of community, 167
   social system, 54
   specialization, 56–60, 62
   speed-or-accuracy trade-off, 124, 169
   stress, 124
   systems approach, 4–5, 53, 61, 164–182, 185–187, 189–190
   team meetings, 3–4, 6–9, 112–113, 178–179
   teamwork, 3–4, 41, 124–125
   training. *See* Training
   work environment, 6, 18
   See also Aviation maintenance; Workplace communication
Airframe & Powerplant (A&P) Mechanic, 56–57, 81, 85
Airline deregulation, 43–45
Airworthiness Directives (ADs), communication regarding, 27, 186
Aloha Airlines accident (Maui, 1988), 21–28, 54–55, 59, 125
"Altitude Awareness" test program (USAirways), 106
AME. *See* Aircraft Maintenance Engineer
AMT-T. *See* Transport Aviation Maintenance Technician

AMTs. *See* Aircraft Maintenance Technicians
ARS. *See* Aviation Repair Specialist
Atlanta accident (ValuJet Airlines, 1995), 67–69, 169
Atlantic Southeast Airlines accident (Carrollton, 1995), 69–71
Authority, cultural redesign and, 186
Aviation industry
   1980s labor problems, 58, 75
   aging fleet, 59–61
   "bathtub curve," 58–59
   deregulation, 43–45
   foreign airlines, 45–46
   human factors management, 40–45
   Reagan-era changes, 58–59
   specialization, 59–60, 62, 63
Aviation maintenance
   1941 CAB creed, 41
   A&P Mechanic designation, 56–57, 81, 85
   Airworthiness Directives (ADs), 27, 186
   AME designation, 79–80
   AMT-T designation, 56, 79, 80, 83, 84
   ARS designation, 84
   "bathtub curve," 58–59
   change needed in, 163–164
   commitment, 83, 85, 87–88, 167
   communication training. *See* Communication training
   competence, 83, 84, 87
   complexity of, 1–3, 4
   control, 83, 85, 87, 167
   deregulation and, 44–45
   forms design, 101–102
   human factors management, 41
   independent repair stations, 65–78
   licensed repairmen, 85
   meetings. *See* Meetings
   new systems approach, 166–182, 185–187, 189–190
   outsourcing, 44
   participative management, 42–43, 45, 114, 171
   performance information, 179
   professionalism, 53–64, 77, 79–88, 173
   quality control, 42, 67–69, 169–170
   Reagan-era changes, 58–59
   recent industry cultural changes, 46–47
   safety as goal, 171, 172
   sense of community, 167
   social system, 54
   specialization, 59–60, 62, 63
   speed-or-accuracy trade-off, 124, 169
   state of maintenance meetings, 178
   statistics, 85–86
   stress in, 124
   systems approach, 4–5, 53, 61, 164–182, 185–187, 189–190
   teamwork, 124–125
   technical training, 181–182
   today's reality, 163–164
   unlicensed mechanics, 65, 67, 85
   See also Aircraft Maintenance Technicians (AMTs); Maintenance program; Maintenance Resource Management (MRM) programs; Workplace communication
Aviation Repair Specialist (ARS) certificate, 84

"Bathtub curve," 58–59
Behavior modification, 54
Beyond Aviation Human Factors (Maurino et al.), 38–39
"Blue-collar blues," 62
Boeing
   Maintenance Error Decision Aid program (MEDA), 145, 147, 180
   Model Maintenance Safety program (MMSP), 144–145, 146, 147
Borman, Frank, 47
Boundary location, principle in cultural redesign, 186
British Caledonia Airlines, participative management, 45

Cambridge cockpit, 41
Carrollton (Ga.) accident (Atlantic Southeast Airlines, 1995), 69–71
Case studies. *See* Accidents
CBI. *See* Computer-based instruction

Index

CCC. *See* Crew Coordination Concepts
Cherns, Albert, 185–187
Civil Aeronautics Act (1938), 43
Cockpit Resource Management (CRM) program, Continental Airlines, 123, 125–126, 127–128
Commitment, 83, 85, 87–88, 167
Communication. *See* Communication skills; Communication training; Workplace communication
Communication skills, as job requirement, 11–12, 182
Communication training, 19, 47, 143
  Boeing MMSP program, 144–145, 146, 147
  computer-based instruction (CBI), 148
  consultants, 182
  Continental Airlines CRM program, 123, 125–126, 127–128
  Continental Airlines MRM program, 123, 126–141, 143–144, 148
  meeting training, 181
  as ongoing program, 165–166, 181
  support for, 182
  Transport Canada HPIM program, 144, 145–146, 147
  USAirways MRM program, 47, 48, 105–122, 147, 166
Compatibility, as principle of cultural redesign, 185
Competence, 83, 84, 87
Complexity, of maintenance work, 1–3, 4
Computer-based instruction (CBI), 148
Computerized log book, 115–116
Conflict, within work teams, 8
Consultants, 176, 182
Continental Airlines
  Cockpit Resource Management (CRM) program, 123, 125–126, 127–128
  communication training, 47
  Crew Coordination Concepts (CCC), 126–127
  Maintenance Resource Management (MRM) program, 123, 126–141, 143–144, 148
  management style, 46, 47

Continental Airlines Maintenance Resource Management (MRM) program, 123, 126–128, 138–141, 143–144, 148
  attitudes and, 130–133
  dependability, 129–130
  goal sharing, 135–137, 140, 156
  management support, 137–138
  MRM/TOQ, 131, 133, 139–140, 151–152
  performance measures, 128
  results, 128–129, 133–135
Continental Express Airlines
  Eagle Lake accident (1991), 21–22, 28–33
  near miss (1992), 33–34
Continuous improvement, cultural redesign and, 187
Control, 83, 85, 87, 124, 167
"Crackdown" approach, 42–43, 45, 179–180
Crew Coordination Concepts (CCC), 126–127
Crew meetings, 112–113, 178–179
CRM. *See* Cockpit Resource Management (CRM) program
Culture. *See* Organizational culture

Decision making
  Boeing MEDA program, 145, 147, 180
  communication quality and, 65–78
  within work teams, 3, 8
Deregulation. *See* Airline deregulation
Direct data entry (DDE), log books, 116
Documentation
  accidents caused by, 68–77
  errors, 15–18, 48, 101
  forms design, 101–102
  of state of the company meetings, 178
  training on, 111
Drury, Prof. Colin, 101

Eagle Lake accident (Continental Express, 1991), 21–22, 28–33
Eastern Airlines
  management style, 46–47
  Miami near miss (1983), 74–75, 165

193

Emery Air Freight, positive reinforcement program, 43
Empowerment, cultural redesign and, 186
Ergonomics, 41, 101
Errors
   Boeing MEDA program, 145, 147, 180
   "crackdown" approach, 42–43, 45, 179–180
   documentation, 15–18, 48, 101
   human factors management, 37, 39, 42–43, 48
   learning from mistakes, 181
   monitoring, 100–101, 179, 180
   outside maintenance vendors, 67
   quality control, 42, 67–69, 169–170
   roundtable discussion, 117–118, 180–181
   specialization and, 63
   systems approach, 180–181
   USAirways MRM program, 106, 116–117
   variance control, 185
   zero-defects programs, 42
   See also Accidents; Maintenance errors

FAR Part 66, 80, 84, 85
Fleet maintenance. *See* Maintenance program
Fleet safety
   Aloha Airlines, 23, 25, 26
   "bathtub curve," 58–59
Flight simulator, 41
Forms design, 101–102, 113–116, 120–121
Functional silos, 140

General Maintenance Manual (GMM), rewritten by USAirways AMTs, 114–115, 121
Goal sharing, at Continental Airlines, 135–137, 140, 156
Goglia, John, 48

Hamilton Standard Company, propeller accident, 70
HPIM. *See* Human Performance in Maintenance program

Human engineering, 41
Human factors management, 27, 38–39, 48–49
   case studies, 46–47
   history, 40–45
   in other countries, 45
   recent aviation industry changes, 47–48
   socio-technical system design, 49
Human Performance in Maintenance (HPIM) program, Transport Canada, 144, 145–146, 147
Human values, cultural redesign and, 187

Incompletion, cultural redesign and, 187
Independent repair stations, 65–78
   case studies, 67–77
   communication, 100
Information flow, cultural redesign and, 186
Information technology, antiquated, 115

Japan Air Lines, participative management, 45
Job card, 76
Joint optimization, 185

Kelleher, Herb, 48

Laissez faire style of management, 55
Lead mechanics, communication needs, 13–14
Leadership, as management function, 174
Letting go, by management, 174
Licensed repairmen, 85
Log book write-ups, 102
Log books
   computerized, 115–116
   design, 113–114, 120
   direct data entry (DDE), 116
   USAirways, 113–116
Looking outward, by management, 174

McDonald's, communication success, 89–90, 92–93
Maintenance. *See* Aircraft Maintenance Technicians (AMTs); Aviation maintenance; Maintenance errors

Maintenance Error Decision Aid program (MEDA, Boeing), 145, 147, 180
Maintenance errors
  Boeing MEDA program, 145, 147, 180
  causes of, 15–18
  communication and, 5
  "dirty dozen," 145–146, 147, 156
  monitoring, 100–101
  posting errors daily, 100
  variance control, 185
  whole-system view, 165–166
  See also Accidents
Maintenance manual, rewritten by USAirways AMTs, 114–115, 121
Maintenance Policies and Procedures, USAirways document, 115, 121
Maintenance program
  Aloha Airlines, 23–26
  NTSB indictment of, 22
  See also Aviation maintenance
Maintenance Resource Management (MRM) programs, 160–161, 165–166, 172
  attitude changes, 153–157, 165–166
  Boeing, 144–145, 146, 147, 180
  changes from, 158–160
  Continental Airlines, 123, 126–141, 143–144, 148
  resistance to, 172
  trainers, 119, 148
  Transport Canada, 144, 145–146, 147
  trends in, 143, 146–152
  USAirways, 47, 48, 105–122, 147, 166, 180–181
Maintenance Resource Management Technical Operations Questionnaire (MRM/TOQ), 131, 133, 139–140, 151–152
Management
  Aloha Airlines accident, 24–27, 54–55, 59
  "bathtub curve," 58–59
  Boeing MMSP program, 144–145, 146, 147
  case studies, 46, 47
  communication by, 16, 18
  Continental Airlines CRM program, 123, 125–126, 127–128
  Continental Airlines MRM program, 123, 126–141, 143–144, 148
  Continental Express accident, 28
  "crackdown" approach, 42–43, 45, 179–180
  deregulation and, 43–45
  empowering staff, 186
  error management, 180
  foreign airlines, 45–46
  human factors science, 42–43
  information flow, 186
  laissez faire style, 55
  leadership role, 174
  letting go, 174–175
  looking outward, 174
  multifunctionalism, 186
  participative management, 42–43, 45, 114, 171
  "people-centered," 180
  professionalism and, 62, 173
  roles of, 174–175
  state of the company meetings, 177–178
  state of maintenance meetings, 178
  state of the site meetings, 178
  support by, 175
  systems culture as new paradigm, 171–182, 185–187
  telephone hotline, 111–112
  top-down communication, 3–4, 5, 17
  Transport Canada HPIM program, 144, 145–146, 147
  USAirways MRM program, 47, 48, 105–122, 147, 166, 180–181
Maui accident (Aloha Airlines, 1988), 21–28, 54–55, 59, 125
Maurino, Daniel, 38, 125
MEDA. *See* Maintenance Error Decision Aid program
Meetings
  performance information, 179
  preshift crew briefings, 112–113
  roundtables, 117–118, 180–181
  state of the company meetings, 177–178
  state of maintenance meetings, 178

Meetings *(continued)*
   state of the site meetings, 178
   team meetings, 3–4, 112–113, 178–179
   training for, 181
   See also Whole-system communication
Miami accident (ValuJet Airlines, 1996), 71–74
Minimal critical specification, as principle of cultural redesign, 185
Mistakes. *See* Maintenance errors
MMSP. *See* Model Maintenance Safety program
Model Maintenance Safety program (MMSP, Boeing), 144–145, 146, 147
Monitoring, maintenance errors, 100–101, 179, 180
Motivation
   behavior modification, 54
   "crackdown" approach, 42–43, 45, 179–180
   positive reinforcement programs, 43
   reward system, 187
MRM. *See* Maintenance Resource Management (MRM) programs
MRM/TOQ. *See* Maintenance Resource Management Technical Operations Questionnaire
Multifunctionalism, cultural redesign and, 186

National Transportation Safety Board (NTSB), report on Maui and Eagle Lake accidents, 22, 23–24, 27, 28, 29
Nonlicensed overhaul mechanics, 67

Organizational culture, 9
   challenging basic assumptions, 164–165
   change needed, 163–164
   changing, 33–34, 118–119, 123–124, 137, 167, 175–176, 185–187
   consultants, using, 176
   new systems culture, 166–182, 185–187, 189–190
   paradigm shift, 165–168
   principles of cultural redesign, 185–187

   resistance to change, 172–173
   transitional organization, 187
   USAirways MRM program, 118–119, 166
   work system redesign, 176–177
Outside vendors, 65–78
Outsourcing, 44

Paperwork. *See* Documentation
Participative management, 42–43, 45, 114, 171
   Boeing MMSP program, 144–145, 146, 147
   Continental Airlines CRM program, 123, 125–126, 127–128
   Continental Airlines MRM program, 123, 126–141, 143–144, 148
   forms redesign, 113–114, 120–121
   new systems culture, 171–172
   Transport Canada HPIM program, 144, 145–146, 147
   USAirways MRM program, 47, 48, 105–122, 147, 166, 180–181
People Express, 44, 46, 47
Performance data, new systems concept, 179
Pilots, selection criteria, 40, 41
Positive reinforcement programs, 43
Power, cultural redesign and, 186
Preshift crew briefings, 112–113
Problem solving, USAirways roundtables, 117–118, 180–181
Professionalism, 53–64, 77, 173
   "blue-collar blues," 62
   communication, 82–87
   control and, 124
   defined, 83–84
   improving, 79–88
   Reagan-era changes, 60–62
   specialization and, 59–60, 62, 63

Quality control
   systems approach, 169–170
   ValuJet Atlanta accident (1995), 67–69, 169
   zero-defects program, 42

Repair stations. *See* Independent repair stations
Reward system, 187
Roundtables, 117–118, 180–181

SabreTech Aviation, ValuJet Miami accident, 71–74
Safety
 Boeing Maintenance Error Decision Aid program (MEDA), 145, 147, 180
 Boeing Model Maintenance Safety program (MMSP), 144–145, 146, 147
 as goal, 171–172
 MRM training, 152
 roundtable discussion, 117–118, 180–181
 training and, 111
 workplace communication and, 1–3, 39, 170–171
 See also Accidents
Scandinavian Airlines, participative management, 45
Sherwood, Jack, 174
Shift turnover, 32, 33
 Continental Express accident (1991), 29–32
 Continental Express near miss (1992), 33–34
 preshift crew briefings, 112–113
 ValuJet Miami accident (1996), 73
Skinner, B.F., 43
Socio-technical system design, 49
Southwest Airlines, 44, 48
Specialization, professionalism and, 56–60, 62, 63
Speed-or-Accuracy Trade-Off (SATO), 124, 169
"Squawks," 102
State of the company meetings, 177–178
State of maintenance meetings, 178
State of the site meetings, 178
Stelly, John, 47
Stress, in aviation maintenance, 124
Stress management, 155, 157
Support system, for cultural redesign, 182, 186–187

Systems approach, 48–49, 164–182, 185–187, 189–190
 aviation maintenance tasks, 53, 61–63, 164–166
 consultants, using, 176
 error management, 179–182
 human factors management, 42–43
 information flow, 186
 management role in, 172–178
 meetings and, 177–182
 multifunctionalism, 186
 need for, 163–164
 organizational objectives, 171–172
 paradigm shift, 164–168
 participative management, 171
 performance data, 179
 power and authority, 186
 principles of cultural redesign, 185–187
 safety as goal, 171, 172
 speed and accuracy, 169–170
 whole-team communication, 4, 6–9, 16, 18
 work systems redesign, 176–177
 workplace meetings, 177–179, 181–182

Taylor, Charles E., 37–38, 163
Team meetings, 3–4, 112–113, 178–179
Teamwork
 in aviation industry, 3–4, 41, 124–125
 Boeing MMSP program, 144–145, 146, 147
 Continental Airlines CRM program, 123, 125–126, 127–128
 Continental Airlines MRM program, 123, 126–141, 143–144, 152
 team meetings, 3–4, 112–113, 178–179
 Transport Canada HPIM program, 144, 145–146, 147
 USAirways MRM program, 47, 48, 105–122, 147, 152, 166, 180–181
Technical bulletins, 75–77, 117
Technician errors. *See* Maintenance errors
Telephone hotline, 111–112
Texas Air Corporation, management style, 46, 47
Top-down communication, 3–4, 5, 17

Training
   Aloha Airlines Maui accident, 27, 55
   of AMTs, 17, 41, 55, 111
   Boeing MMSP program, 144–145, 146, 147
   computer-based instruction (CBI), 148
   Continental Airlines CRM program, 123, 125–126, 127–128
   Continental Airlines MRM program, 123, 126–141, 143–144, 148
   Crew Coordination Concepts (CCC), 126–127
   Eastern Airlines near miss (1983), 75–76, 165
   error management, 180
   meeting training, 181
   for paperwork, 17
   for "people-centered" management, 180
   of pilots, 40, 41
   problem solving, 181
   safety training, 111, 125–126
   technical training, 181–182
   Transport Canada HPIM program, 144, 145–146, 147
   USAirways MRM program, 118–119, 166
   See also Communication training
Transitional organizations, cultural redesign and, 187
Transport Aviation Maintenance Technician (AMT-T), 56, 79, 80, 83, 84
Transport Canada, Human Performance in Maintenance (HPIM) program, 144, 145–146, 147
Turk Hava Yollari (THY), ValuJet Airlines engine accident (1995), 67–69
TWA, improved communication at, 48

United Airlines, improved communication at, 47–48
Unlicensed mechanics, 65, 67, 85
USAirways
   "Altitude Awareness" test program, 106
   audit productivity, 107
   improved communication at, 48, 166

   MRM program, 47, 48, 105–122, 147, 166
   signatory requirements, 108
   use of stamps on forms, 108, 120
USAirways Maintenance Resource Management (MRM) program, 47, 48, 105–106, 119–122, 147
   computerized logbooks, 115–116
   error patterns, 116
   history, 106–107
   initial activities, 108–111
   maintenance manual rewritten, 114–115
   management support, 119–120, 166
   paperwork design, 113–114
   procedural changes from, 109–113
   roundtables, 117–118, 180–181
   training issues, 118–119

ValuJet Airlines
   Atlanta accident (1995), 67–69, 169
   Miami accident (1996), 71–74
Variance control, as principle of cultural redesign, 185
Vietnam War, human factors management, 42

Whole-system communication, 4, 6–9, 16, 18
Work cards, form design, 101–102
Workplace communication
   accidents resulting from, 2, 21–33. *See also* Accidents
   Airworthiness Directives (ADs), 27, 186
   as cause of mistakes, 5, 15
   cost pressures and, 65–78
   data transfer, 3, 81
   decision-making and, 3, 8, 65–78
   defined, 1
   as dialog, 3
   human factors management, 37–46
   independent repair stations, 65–78
   information flow, 186
   management and, 16, 18
   meetings. *See* Meetings
   mission communication, 170

Workplace communication *(continued)*
  MRM programs. *See* Maintenance Resource Management (MRM) programs
  multimedia nature of, 89–93
  NTSB indictment of, 22
  oral communication, 93–94
  outside vendors, 65–78
  preshift crew briefings, 112–113
  principles for cultural redesign, 185–187
  professionalism and, 53, 63, 77, 81–82
  Reagan-era changes, 60–61
  roundtables, 117–118, 180–181
  safety and, 1–3
  telephone hotline, 111–112
  top-down communication, 3–4, 5, 17
  "whole-system" view, 4, 6–9, 16, 18, 53, 61–63, 164–166
  written communication, 92, 94–103
  See also Communication skills; Communication training
World War I, human factors management, 40
World War II, human factors management, 40–41
Wright-Martin Company, 37
Write-ups, 102
Written communication, 92, 94–103

Zero-defects program, 42

About the Authors

James C. Taylor, Ph.D. Jim is Adjunct Professor in the Engineering Department at Santa Clara University in California and he is the principal in his own company, Sociotechnical Design Consultants. His international research and consulting experience, including work in the aviation industry, spans the last 30 years. Jim has authored or co-authored six books on organization behavior and work system design and has published another 70 papers on these topics.

His longtime theme in teaching, research, and consulting is to improve both quality of working life and organizational effectiveness through the design of work systems.

T.D. Christensen. Tom is president of STS Organization Consultants in South Bend, Indiana. He is an award-winning business communicator, author of educational materials used by trainers on four continents, and a consultant in the design and implementation of team-based work systems and total communication programs.

He is especially interested in the ways that physical material, information, and people are all transformed by their interaction as they pass through the workplace.